EIGHT PREPOSTEROUS PROPOSITIONS

EIGHT PREPOSTEROUS PROPOSITIONS

From the Genetics of Homosexuality

to the Benefits of Global Warming

Robert Ehrlich

PRINCETON UNIVERSITY PRESS

PRINCETON AND OXFORD

Copyright © 2003 by Princeton University Press
Published by Princeton University Press, 41 William Street,
Princeton, New Jersey 08540
In the United Kingdom: Princeton University Press, 3 Market Place,
Woodstock, Oxfordshire OX20 1SY
All Rights Reserved

Second printing, and first paperback printing, 2005
Paperback ISBN-13: 978-0-691-12404-9
Paperback ISBN-10: 0-691-12404-3

The Library of Congress has cataloged the cloth edition of this book
as follows

Ehrlich, Robert, 1938–
Eight preposterous propositions : from the genetics of homosexuality to
the benefits of global warming / Robert Ehrlich.
p. cm.
Includes bibliographical references and index.
ISBN-13: 978-0-691-09999-6 (acid-free paper)
ISBN-10: 0-691-09999-5 (acid-free paper)
1. Science. 2. Hypothesis. I. Title.

Q171.E3735 2003
500—dc21   2003043328

British Library Cataloging-in-Publication Data is available

This book has been composed in Palatino Typeface

Printed on acid-free paper. ∞

pup.princeton.edu

Printed in the United States of America

10   9   8   7   6   5   4   3   2

This book is dedicated to my sisters, **Millie and Linda**

Contents

Acknowledgments

I AM GRATEFUL to the many people who provided feedback on selected chapters of this book, including Michael Behe, Gerald Bracey, William Byne, David Ehrlich, Robert Ellsworth, James Flynn, Dean Hamer, Arthur Hobson, Robert Jahn, Menas Kafatos, David Lasso, John LaRosa, Patrick Michaels, Harold Morowitz, Dean Radin, Uffe Ravnskov, Shobita Satyapal, Jagadish Shukla, Robert Temple, Albert Torzilli, and John Wallin. A number of colleagues provided feedback on multiple chapters, including Len Feingold, Harold Geller, Robert Karlson, Eugenie Mielczarek, and Michael Summers. And some reviewers were kind enough to provide feedback on the entire manuscript, including Robert Fudali, Joel Goldstein, Edmund Habib, Mitchell Hobish, Dara Koper, Selden McCabe, Fred Singer, and Kichoon Yang. Any errors in the book are, of course, entirely my responsibility.

WE LIVE IN AN AGE when the boundaries between science and science fiction are becoming increasingly blurred. It sometimes seems that nothing is too strange to be true. How can we decide which of the outlandish ideas that are constantly bombarding us by might be true and which are complete nonsense? It's particularly tough to come to an informed judgment regarding ideas that have a scientific component where there are also huge political and economic stakes involved, such as global warming. In such a case, it's too easy to go by our political preferences or by the opinions of the last expert we saw on TV. This book will help you look at the "evidence" critically and judge for yourself. It is similar to *Nine Crazy Ideas in Science—A Few Might Even Be True*. In this book, though, we'll be looking at topics that are a bit less "sciency" and closer to everyday life. But we will still be using the tools of science to analyze each topic.

Scientists try to take as little for granted as possible. They are constantly asking how we know whether something is true, and whether alternative explanations are possible. Scientists also seem to be more comfortable with uncertainty than other people. Many scientists may believe that a given theory—say, that of human-caused global warming—is likely to be true, but be far from certain about it. Such uncertainty is disquieting to most people who want to have as clear a picture as possible of our planet's future. Policy-makers, in particular, want informed scientific guidance before undertaking expensive solutions to problems of uncertain severity.

People facing large uncertainties and potentially large risks may be forced to rely on their gut instincts. That's OK in one sense. Making judgments based on our emotions is under-

2   standable if our large uncertainties reflect those that are inherent in an issue. But if the uncertainties are simply a matter of our own laziness in not looking into an issue deeply enough, then relying on our instincts is very unfortunate. Public policy on important issues may be left by default to those competing interests that are best able to manipulate public opinion.

Scientists approach a controversial theory from the opposite perspective of lawyers. Lawyers are paid to make the best case possible for their client, a person they may believe to be guilty. They want to know all the evidence on the other side, but only in order to refute it better and make their case stronger. If they can't refute the evidence, they'll try any trick in the book to make the jury discount it. Scientists, on the other hand, are more like jury members. While science does include its share of cheaters, self-deceivers, and self-promoters, scientists should judge a theory by fairly weighing all the evidence on each side.

Like juries, people who approach topics from a scientific perspective also may require different levels of certainty ("beyond a reasonable doubt" or the less strict "preponderance of the evidence"), depending on what the theory involves. I suspect that on global warming, for example, the preponderance of the evidence might be enough certainty for most people. But, unlike a jury, which has to make up its mind once and for all, scientists must continue to remain open-minded to contrary evidence even after they have accepted a theory as being probably true.

Why did I choose this particular set of eight topics? As in *Nine Crazy Ideas*, I wanted to explore some topics that are controversial, and I hope that I've included enough controversial ones to offend just about everyone. I also wanted topics that have important public policy implications in the real world and have ample scientific data on each side of the issue. Although initially I was reluctant to tackle any topics involving paranormal or psychic phenomena, I wound up including one (the possibility that objects can be influenced by thought alone) because a great deal of experimental data actually exists on this matter.

Although in many cases I found the evidence for an idea not to be credible, this is not a debunking book. I tried to look at

3   each idea as objectively as possible. Even though I did have some initial biases with most of the eight ideas, I also found myself changing my opinion on them—in some cases two or three times! As a general rule, I believe that in approaching a new idea it is very important not to make up your mind too quickly.

Is there a formula to apply to a given controversial idea to see if it might be true? No, but you should ask yourself questions: How do the proponents of a new idea claim to know that it's true? How might the data be interpreted differently? How can the theory be tested? Let me give you an example from a topic that I almost decided to write about, but didn't. Many people believe that violence in the media causes children who view it to become violent. There is no question that a connection exists at some level between media and real-world violence, given the copycat phenomenon. Even some terrorists are said to have gotten ideas from viewing action–adventure movies! But here we're considering the more general question of whether media violence causes children exposed to it to become violent later in life. Not having looked at the research, I am agnostic on the question of whether this view is correct. Most social scientists believe it is true, but there are some who don't.[1] What are some of the kinds of questions you might ask yourself to decide whether the theory that media violence causes real life violence is true? Why don't you actually stop reading for a few minutes and make a list. You could then later compare your list to mine at the end of this chapter.

In evaluating a controversial idea it is useful to consult a wide range of sources. The Internet makes this quite easy, but it also makes it easy to wind up at web sites of pseudo-experts. One very helpful summary of methods to check that the information on a web site is reliable has been prepared by Jim Kapoun, an instruction librarian at Southwest State University—see his site at www.ala.org/acrl/undwebev.html. Kapoun's checklist was prepared to help college students evaluate the reliability of any web site, but its methods are appropriate for anyone. You can hone your skills at recognizing the real and pseudo-experts on

4   any topic by answering the specific questions that Kapoun raises about any web site. Here are some additional criteria you can use. More often than not, the pseudo-expert's sites share these features. (1) Pseudo-experts are usually certain of everything they claim, (2) cite their own research frequently, (3) try to impress you with fancy titles, (4) describe the suppression of their ideas by the establishment, and (5) have a clear agenda and maybe even a financial incentive. But aside from the pitfalls of being swayed by fancy-looking web sites that actually contain nonsense, the web is a fantastic tool for gaining valid information on virtually any topic.

It may strike you that I am being hypocritical in stressing the importance of being wary of pseudo-experts on any given topic. After all, am I—a physics professor—not just a pseudo-expert on many of the topics I'm writing about in this book? For that matter, how can you or I be expected to find the truth about controversial subjects that lie way outside our fields of expertise, when even the real experts disagree? I think it *is* actually possible for outsiders to do a competent job of analyzing evidence in many areas—but only if they have done their homework.

That homework involves learning a bit of statistics (the favorite tool of people who want to distort the truth) and enough of the basic science and vocabulary in a field to clearly understand the basis for what is being claimed. As I've already said, just follow the evidence, ask how each claim is demonstrated to be true, and don't make up your mind too quickly. If you decide too quickly, you could fall into the trap of filtering all evidence through your preconceived view and not giving contrary evidence sufficient weight—which we all do far too frequently. That trap is the very essence of prejudice.

The eight topics discussed in this book have varying degrees of credibility. In each case, after discussing the evidence on each side I give the idea a rating at the end of the chapter, according to how well the case for it has been demonstrated. In a previous book I used a "cuckoo" rating scale, which went from zero to four cuckoos, based on how crazy I considered an idea. Here I've abandoned the cuckoo scale in favor of a "flakiness" scale. An example may help clarify the difference between

5   craziness and flakiness. Many of the ideas of modern physics are completely crazy, especially some of the paradoxical ideas of quantum physics. In fact, when one of the pioneers in this field, Wolfgang Pauli, made a presentation on one of his new crazy theories, he was told by the great physicist Niels Bohr: "We are all agreed that your theory is crazy. The question, which divides us, is whether it is crazy *enough* to have a chance of being correct. My own feeling is that it is not crazy enough."

Bohr understood that great revolutionary advances in science always will sound crazy at the beginning, partly because they challenge conventional wisdom, and partly because they will initially be presented in a confusing incomplete form. The tidying-up done after the fact makes great scientific advances seem much more logical, but it obscures the important roles of intuition, imagination, chance, and even aesthetics in making scientific discoveries. The true scientific method is not quite as neat and logical as we sometimes portray it to be.

Although many of the ideas of quantum physics are still crazy after all these years (since the 1930s), they are certainly not "flaky," i.e., lacking in empirical evidence or internal consistency. Conversely, new theories may be flaky, but not sound crazy at all, if they fit into your view of how the world works. My new rating scheme based on flakiness goes from zero flakes, meaning a reasonable degree of confidence that the idea is true based on the evidence, to four flakes, meaning no credible evidence for the idea. A summary of my ratings for each idea can be found in the epilogue. Obviously, these ratings are subjective and influenced by my own biases, but I will reveal those biases if I have any, so that you can decide for yourself how honestly I've dealt with each idea.

Questions to Ask in Judging Whether A Really Causes B

This section illustrates some questions you might ask to decide whether a theory claiming that A causes B is well supported by the evidence. For specificity, we'll assume that A is media violence to which children are exposed and B is real-life violence

6   that they later commit, but the same questions could serve as a template for just about any other topic. I suspect you may be able to come up with a number of additional questions to those listed.

1. How exactly do the studies looking for a connection between A and B define and measure A? How do they define and measure B? Do their definitions seem reasonable?
2. Have studies shown there is a correlation between A and B? How strong is the correlation?[2]
3. If only some of the studies on A and B show there is a correlation between them, which studies seem better designed? Which studies have the greater statistical significance? (Just because a study finds a negative result, it doesn't mean A and B are unrelated—for example, the study sample may have been very small.

Statistical Significance

This is the first of many boxes throughout the book. Boxes contain information that can be skipped without disturbing the flow. Sometimes the information in the box may be a little more technical than the text. Other times it may provide definitions or examples of ideas discussed in the chapter. This box concerns the important subject of statistical significance. The statistical significance of any study is judged by the likelihood that the results could have been due to chance. Researchers in different fields have different standards regarding how small this likelihood should be for a result to be considered "significant." In many fields the criterion is "95 percent confidence," meaning that there is a 5 percent probability (often written $p < 0.05$) that the result could be due to chance, and therefore bogus. Clearly, the smaller the probability or p value, the higher the level of statistical significance and the more confidence we can have in a study's results. The simplest way that a study can achieve a higher level of statistical significance is to amass more data, which is sometimes costly or not feasible.

---

4. If some studies show that there is a strong correlation between A and B, can that correlation be explained in ways other than A being the *cause* of B?[3]

5. How do the studies exploring the connection between A and B control for confounding variables, such as other possible causes of B? What are those other causes, and do the studies investigate how important they are compared to A?

6. Do the studies compare situations in which A is present with "control groups" in which A is not present? If so, how can we be sure the control groups are representative?

7. Do the studies show that the relation between A and B is a *continuous* one, that is, the more A is present, the more B is found later?

8. If A really causes B, can we explain why in some places A is common but B is not? (Japan, for example, has a great deal of media violence, but little real-life violence.)

9. If A really causes B, can we explain why A becomes at some times more prevalent while B becomes less prevalent? (Media violence has probably been increasing in the United States over time, but real-life violent crime has been decreasing for a number of years, at least until 2001.)

10. Do the individuals doing studies on the connection between A and B appear to have strong ideological biases?

12. Who is funding a given researcher's study?[4]

I VIVIDLY remember the night that I declined to play on the other team. I was a seventeen-year-old virgin employed as a lifeguard at a summer resort in the Pocono mountains of Pennsylvania. Lenny, a good-looking male dancer, had just propositioned me. Lenny correctly pointed out that I couldn't properly evaluate gay sex if I hadn't tried it, so why not give it a try? His offer did have a certain allure—especially since I hadn't yet had any luck in my attempts to get young women to go to bed with me. But "no thanks," I told Lenny. "It isn't my thing." Had I, in fact, taken Lenny up on his offer, I would have been one of the estimated 16 to 37 percent of American males who have had a homosexual experience, most of whom are not gay.[1] But can young heterosexual men be seduced into the homosexual lifestyle by such experiences, or are these men simply gay without being fully aware of it?

Those readers who believe that homosexuality is not innate will, of course, answer that gay seduction is a real possibility—especially in our current society, which has destigmatized homosexuality and even made it glamorous in the eyes of some teenagers. Worries about young people being lured into a homosexual lifestyle are one factor behind heterosexual American reservations about gays taking on roles such as parents, clergy, scout leaders, and teachers. It is unfair to characterize all (of the roughly 40 percent of Americans) who hold such views as ignorant bigots. Even if they are aware of studies that show that homosexuals are no more likely to be child molesters than heterosexuals,[2] some people may still worry that children who interact with gays may somehow be attracted to the gay lifestyle.

9   On the other hand, those who view sexual orientation as an immutable characteristic of a person are likely to not worry about seduction into the gay lifestyle, and to view gays in a more favorable light. They might argue that homosexuality is no more a matter of conscious choice than heterosexuality. Of course, I did make a conscious choice not to take Lenny up on his offer, but that is not the choice that I am discussing here. Rather, I have no recollection of making a choice to be a heterosexual in the first place, just as most gays claim not to have any awareness of making a choice to be gay.

Given the pressures in our society in favor of heterosexuality, many gays probably have had "reverse-Lenny" experiences. Having tried or at least considered sex with members of the opposite sex, they may have concluded after some time that it wasn't for them. In fact, it appears that some homosexuals make such a discovery only after being married for some years to a member of the opposite sex. As with heterosexuals, however, most gays' awareness of their sexual orientation probably occurred to them long before their first non-autoerotic sexual experience. (I grew up fantasizing about Marilyn Monroe long before I met Lenny.)

In fact, even though sexual orientation is not usually assessed before puberty, many researchers believe that our sexual orientation is largely determined at a very early age— perhaps as young as two. One reason for this belief is the studies of young children, which have shown that 80 percent of youngsters who show extreme gender nonconformist behavior turn out to be gay, lesbian, or bisexual.[3] (Examples of such behavior among boys would include playing with dolls persistently, cross dressing, or otherwise acting effeminately, and avoiding rough and tumble play). The true figure may even be higher than 80 percent because the latest interviews in the study were held when the children had reached age 18, and some may not yet have acknowledged their homosexuality. (Note that the preceding statistic does not mean that 80 percent of homosexuals showed gender nonconform-

10 ist behavior as children—in fact, about two-thirds showed such behavior.[4])

A skeptic could argue that when gays say that they did not choose their lifestyle they are being dishonest. Given the political stakes involved, they would have every incentive to pretend that their homosexuality was not a matter of choice, even if it were. The "born that way" scenario would probably make people more sympathetic to gays. More importantly, it could also be used as a strong justification to give gays the same legal protection as other classes of citizens. If their status as gays is as immutable as that of race, for example, it would—if other specific conditions are met—place them into a so-called suspect class, which is a legal term for groups of people who are entitled to extra protection.

The courts carefully scrutinize laws that affect people in suspect classes, such as race or national origin, to be sure that if the law has a disparate impact on people in such classes, it must be for a "compelling state interest." On the other hand, even if heterosexuals are unwilling to accept gay beliefs about the origin of their sexual orientation, they do have their own lives to look back on. If you happen to be heterosexual, do you recall ever making any conscious choice to be that way?

But perhaps heterosexuality doesn't have to be chosen, because it is the "natural" or default sexual orientation that everyone is born with, unless something goes "wrong." Even so, we could still ask, what is that something, and was its "going wrong" a matter of conscious choice or upbringing? In other words, just how is homosexuality determined? Is it genetic or based on environmental factors (or some combination of the two), and how many of those environmental factors represent choices?

It should be clear by now that in referring to sexual orientation, we are not referring simply to behavior, but to the sex to which a person is attracted. Alfred Kinsey, the pioneering sex researcher, used a seven-point scale ranging from 0 to 6 to classify the sexual orientation, where 0 represented completely heterosexual, and 6 represented completely homosexual.[5] Figure

11   2.1 shows the percentages of people Kinsey reported whose sexual orientation score corresponded to various numbers on his seven-point scale. Kinsey's criteria for scoring sexual orientation include

1. *Attraction*—To which gender are you are sexually attracted?
2. *Fantasy*—Which gender do you fantasize about during masturbation or dreams?
3. *Self-identity*—What orientation do you consider yourself to have?
4. *Experiences*—With which gender have you had sex?
5. *Arousal*—What is the gender of people in erotic images that cause you to become measurably aroused? (Arousal is evaluated for men using a so-called plethysmograph, which fits around the penis and measures partial erections. I'm currently working on a version, like a police radar gun, that you could simply point at people. Just kidding!)

If you want to find out where you belong on the Kinsey scale, or a refined version developed by Fritz Klein, check out the

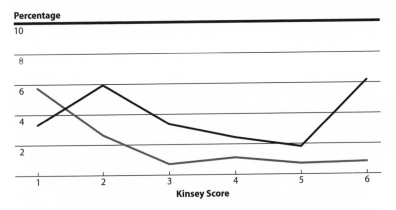

Figure 2.1 Sexual orientation distribution. Percentage of men (dark, top) and women (light, bottom) according to their sexual orientation, using Kinsey's scale and his data from 1948 and 1953.[5] Only Kinsey scores greater than zero (pure heterosexual) have been shown.

12   web site www.biresource.org/pamphlets/klein_graph.htm. At that site, you can find a description of the scale and instructions for seeing where you fit on it. Clearly, some of the above five Kinsey and Klein criteria would appear to be more reliable than others in revealing a person's sexual orientation. Arousal, for example, doesn't depend on truthfulness and is therefore probably the most reliable, since there are various reasons why people might wish to hide their sexual orientation (or simply fail to recognize it). It is reported, for example, that some homophobic men when tested with a plethysmograph are aroused by gay pornography, confirming the long-held belief that for some people their homophobia is based on their own feared latent homosexuality.[6]

This theory of homophobia was proposed by Sigmund Freud, who also expounded the view that male homosexuality was created in childhood as a result of a distant rejecting father and a domineering mother. Thus, Freud certainly did not adhere to the "born that way" view. Although Freud viewed homosexuality as a disorder, he did acknowledge that homosexuality was very difficult to change once it was established, and he had reasonably positive attitudes toward gays for his time. Homosexuality is no longer considered a disorder—having been removed from the American Psychiatric Association (APA) list of mental disorders in 1973—but the view that it results in part from parental–child interactions in early childhood has not been abandoned. The "gender identity disorder (GID) of childhood" is still listed as a mental disorder by the APA.[7] (Actually, while gender nonconformist behavior is a predictor for later homosexuality, gender identity disorder is a much more extreme affair that includes kids who are depressed, suicidal, and insistent they will grow up to be the other sex. While many gays may have exhibited gender nonconformist behavior as kids, few probably had GID.)

Earlier we noted that gender nonconformist behavior in kids is a good predictor of later homosexuality. Because of the early age that such behavior becomes apparent the possibility that the roots of homosexuality are innate becomes more plausible.

13 But what could those roots be and are they the basis for homosexuality itself or merely for the behaviors or traits sometimes associated with it, such as being effeminate? In other words, could it be that effeminate boys are more likely to turn out gay solely because of the reactions of others to their behavior?

Is Homosexuality Natural, Innate, and Immutable?

To be clear about the various issues concerning sexual orientation, let us carefully distinguish between terms such as natural, innate, immutable, biological, heritable, and genetic, and then see how well these terms apply to homosexuality. One definition of the term "natural" means consistent with what is found in nature. Since many species of animals, including the bonobo chimpanzees, our closest relatives genetically, engage in homosexual behavior, it could be argued that homosexuality is in some sense "natural." Of course, on that same basis, one would

Figure 2.2 Wendell and Cass, two "gay" penguins at the New York aquarium. Photo by Rick Miller, New York aquarium, printed with permission.

14 have to admit that rape, cannibalism, and even murder are also natural. Thus, being natural doesn't convey any special degree of acceptability in a moral sense.

It has been suggested that the "homosexual" acts exhibited by many animals are primarily for purposes of establishing dominance rather than expressions of true sexual desire. But that explanation doesn't account for the many observed instances of mutual or reciprocal sexual activity. In any case, the view that all sexual expression must naturally relate only to reproductive activity flies in the face of our observations of the animal kingdom. For that reason it seems justifiable to refer to homosexuality as being "natural." Of course, you are entitled, if you wish, to define all sex for purposes other than reproduction as being "unnatural," in which case homosexuality is not natural, *by definition*. But, in that case, many heterosexuals (including former presidents) have also routinely engaged in unnatural sexual activity, which is no more unnatural (except by biblical fiat) than that engaged in by homosexuals.

A trait is innate if we are born with it, and it is immutable if it cannot be changed, except by medical intervention. With respect to anatomically observable characteristics (such as sex and eye color), it is usually obvious that we are born with them. However, with more complex behavioral-related characteristics (such as left/right-handedness, musical ability, and sexual orientation), evidence on whether we are born with the property is less direct. Usually, all that can be said with certainty is that we are born with a certain tendency, and the strength of that tendency can range anywhere from slight to very pronounced, depending on the property and the individual.

Now innateness and immutability do not always go hand in hand. An example from the animal kingdom is bird songs. According to ornithologist Evan Balaban, most species of birds do not have to be taught their song by their parents.[8] They know it from birth and the song never changes during their lives. However, there are some species of birds, such as robins, who must hear their song from another robin to be able to sing it, and they must hear it within a precisely defined time period

15   after birth. If they happen to hear another bird species song first before their own, they will sing that song for their entire life. For robins, the ability to sing the way most other robins sing is therefore immutable, but it is learned and not innate.

Research on whether human traits are innate and/or immutable seems to be either innocuous or highly controversial, depending on the trait. For example, few people would likely feel it necessary to attack findings that left-handedness seems to be an innate trait, but many more would likely be upset by research that made the same claim about homosexuality. The analogy between these two properties (handedness and sexual orientation) is an interesting one because they appear to have so much in common, as pointed out in Chandler Burr's excellent book *A Separate Creation*.[9]

A Comparison between Handedness and Sexual Orientation

As Burr points out, all of the following facts appear to be true for both traits.

1. *Both handedness and sexual orientation are bimodally distributed.* This means that both traits tend to peak at their two extreme forms. For example, there are few truly ambidextrous people. If we look at how people are distributed according to the Kinsey sexual orientation scale, we find peaks at 0 and 6, with smaller numbers at 1 through 5—at least in the case of men (see figure 2.1). Females show a greater frequency of bisexuality than males, and studies have found that, unlike males, they sometimes change their sexual orientation a number of times during their lives. According to Kinsey's 1948 data, there are (were) extremely few complete lesbians (0.8 percent), which is probably an underestimate, based on more recent surveys. Despite the bimodal nature of the male sexual orientation distribution, Kinsey himself stressed that this human trait is not "one way or the other," but that people's sexual orientation falls on a continuum. Indeed, as his data show, a significant

16   percentage of males (16.7 percent) do have an orientation that is intermediate between completely heterosexual and completely homosexual. (A "mostly" heterosexual with a Kinsey score of 1 might be unlikely to acknowledge that he/she is anything but heterosexual.)

2. *Fewer than 20 percent of the population shows the minority orientation.* The figure for left-handedness is around 8 percent, but obtaining an accurate figure for homosexuality is more problematic. Typically, surveys find that somewhere between 4 and 17 percent of people have a homosexual orientation.[10] It is also difficult to tell how these percentages may have changed over time since Kinsey's report, because the societal climate toward homosexuality greatly influences how honestly people respond to surveys. When surveys of highly personal and controversial questions yield different results, we need to consider the effects of survey bias and interviewee deception. It is understandable that some homosexuals might not wish to reveal their sexual orientation, especially in a less tolerant era.

3. *The minority orientation occurs more often in males.* For left handedness, the incidence is about 30 percent higher in males, while many studies claim that homesexuality is twice as common in males.

4. *The minority orientation is universal.* Both left-handedness and homosexuality occur in all races and cultures, and have been present since antiquity. When Kinsey's groundbreaking study of human sexuality first came out in 1948, it was a great shock to many Americans to learn that homosexuality was present throughout the country, from rural areas to large cities, and in all socioeconomic groups and ages. It is difficult to know whether the frequency of either left-handedness or homosexuality is the same in all cultures and races, however. Some cultures show much less acceptance for homosexuality (and others for left-handedness), and surveys would underreport the true incidence in such societies. One 1993 study of homosexual-

19   natural—became gay.[13] This figure is slightly higher than the most common estimates for the population generally, which could indicate some small degree of environmental influence. But it could also indicate (1) that kids raised by gay parents are more willing to admit it if they happen to turn out gay, or (2) that kids raised by a biological parent, who later came out as gay, inherited the parent's "gay genes.")

11. *Twin studies*. The results of studies of twins are somewhat controversial, for reasons we will discuss later. However, the chances of one twin being left-handed or gay are greater if the other is of that orientation. As would be expected for an inherited trait, the chances are especially increased for identical twins (who have the same genes), as compared to fraternal twins who share 50 percent of their genes, just like other siblings.

Despite the similarities between the traits of handedness and sexual orientation, there is one obvious difference between them. The difference is so stark that it appears to justify labeling the notion that sexual orientation is inherited a "crazy" idea. For an inherited trait to remain in the gene pool over many generations with a frequency approaching 5 or 10 percent, it must convey some benefit in terms of reproductive success, or at least not be a significant detriment. Your right/left-handedness apparently conveys neither a strong benefit nor a detriment in terms of passing your genes on to the next generation. Lefties may be more susceptible to a variety of illnesses and have a shorter life span than righties, but most of them live long enough to breed, and that's all that counts in evolution.

Being homosexual, however, would obviously lower reproductive success much more than being left-handed. Far fewer gays have children than nongays. You might think, therefore, that if a mutation ever did result in creating a gay gene, evolution (which rewards reproductive success) would shortly remove it from the gene pool. On this basis, the idea of a gay gene seems just about as crazy as a gene for celibacy. Strangely, there are several ways that evolution might operate to favor a gay

20   gene. We shall discuss these after considering the evidence that such a gene might actually exist.

The preceding comparison between homosexuality and left-handedness suggests that both traits probably are largely innate, rather than learned. In the remainder of this chapter we shall examine what kind of biological basis homosexuality might have. ("Biological" is a broader category than "genetic," since it includes observable changes in the body that could result from environmental rather than genetic influences. "Environmental" influences can include prenatal ones.) The three types of evidence we shall consider are hormonal influences, brain structure differences, and the possibility of a gay gene.

Hormonal Influences on Sexual Orientation

The sex hormones, including the male hormone testosterone and the female hormone estrogen, have a great influence on our degree of masculinity or femininity. Stereotypically, many gays have feminine personality traits and many lesbians have masculine traits, which suggests that perhaps hormones may have something to do with the development of homosexuality. (Of course, many gays and lesbians defy the preceding stereotypes and, conversely, some heterosexuals fit them—to the point where even gays probably cannot count on their "gaydar-detector" operating at much greater than chance levels.)

That there is a connection between sex hormones and the development of sexual orientation seems plausible, and this idea was an early candidate theory as to how homosexuality develops. This theory was shown to be incorrect when it was found that homosexuals and heterosexuals have, on average, comparable levels of sex hormones. Such a finding should hardly have been a surprise, since sex hormones also influence observable secondary sex characteristics (such as the growth of facial hair and breasts), and there are usually no obvious physical differences of this kind between the two sexual orientations, although some studies have reported differences.

21   Even if adult homosexuals have roughly the same hormone levels as heterosexuals, there remains the possibility that our prenatal levels of sex hormones could have determined our sexual orientation. The evidence that prenatal hormone levels may strongly influence our sexual orientation comes largely from animal studies, since the relevant experiments cannot be ethically carried out in humans. Experiments have been done with many species of mammals, including rats and guinea pigs, in which male fetuses are deprived of testosterone and female fetuses are given more testosterone than usual. The result is that the male animals are "feminized" and the female animals "masculinized" in terms of their later mating behavior as adults.

Specifically, feminized male rats raise their rump (exhibit "lordosis"), permitting other male rats to mount them, and masculinized females mount other females. Other experiments show that it is not just the animal's sexual *behavior* that gets reversed, but also the sex to which they are attracted when the animals are given a choice of male or female mates. When male ferrets were given a choice between a sexually receptive female and another male in two arms of a T-shaped maze, about 4 percent of the male ferrets chose the male over the female.[14] Four percent is not a large number, but that same 4 percent of ferrets chose males consistently.

I'm not aware of any experiments that evaluate the response of male rats or ferrets to erotic images using a plethysmograph. However, the previous results indicate that true homosexuality exists in (1) animals whose prenatal hormone levels have been manipulated, and (2) a small percentage of unmanipulated animals as well. Just as in the case of humans, the experiments showed that the "homosexual" rats tended to demonstrate cross-gender behavior (including aggression level) when they were young.

Although we cannot carry out similar experiments with humans, nature has in some cases done the experiment for us. For example, there is a disorder known as congenital adrenal hyperplasia (CAH), in which female fetuses are exposed to higher levels of male hormones than are normal. During early childhood,

22   CAH girls tend to exhibit cross-gender behavior, such as playing with trucks rather than dolls, more often than other girls.[15]
They also tend to score higher on such "male" abilities as spatial
relations. Finally, CAH girls are more likely to be lesbians than
other girls when they grow up. Although a majority of CAH
girls, in fact, do *not* later become lesbians, that might mean only
the male hormone levels did not become high enough during
some critical prenatal period. One basic conclusion seems to follow from all the hormone research in animals and humans: the
degree of brain (and anatomical) masculinization that occurs in
males and females depends on the level of male hormones to
which the fetus is exposed in some critical period.

  Our sex (male or female) is not simply a function of our
anatomy and genes. Our brains are also influenced by sex hormones. In most people, the "brain sex" and "anatomical sex"
match one another. But it has been suggested that we can think
of homosexuals as being a third (and fourth?) sex, whose brains
are partly wired corresponding to the opposite anatomical sex.
This idea, first suggested by the physician and sexologist
Magnus Hirschfeld, may seem to be a more appropriate model
for transsexuals (people convinced they are trapped in a body
of the other sex) than homosexuals.

  On the other hand, since homosexuals are roughly 2000
times more common than transsexuals, we might speculate
that the same hormonal process occurred in both cases, but that
it was carried to its extreme in the case of transsexuals. Even
though the idea of homosexuals having the brain of the opposite sex is a highly simplistic one, it does fit the observation that
many traits of homosexuals are partly shifted toward the other
sex. (Studies have shown, for example, that gay men tend to engage in less violent behavior than others of their sex, while for
lesbians and bisexual women the reverse pattern was found.[16])

Brain Studies

  If we crudely think of homosexuals as having brains that are
more similar to those of the opposite sex than their own, it is

23   natural to examine brain differences based on sex as a clue to
what to expect for differences due to sexual orientation. (It
would be highly surprising if there were no sex differences in
the human brain, given that sex differences appear in many or-
gans of the human body.) One obvious difference is that male
brains tend to be slightly larger.[17] What about differences in
particular brain structures?

Two brain structures, known as the anterior commissure and
the corpus callosum, were at one time claimed to have different
sizes in men and women. However, attempts to replicate those
claims gave conflicting or negative results.[18] One brain structure that would seem to be a promising candidate for studies
on sex and sexual orientation differentiation is the hypothalmus—a small region near the pituitary gland that regulates
body temperature and controls various drives, including hun-
ger, thirst, aggression, and sex. Studies of rodents have shown a
clear difference according to sex in the size of some structures
in the hypothalmus. For example, one group led by Roger
Gorski found in 1977 that a certain group of hypothalmic cells
is five times larger in male rats than in female rats.[19] This sex
difference has been found in some other studies and in a variety of other animal species, but not all. (Unfortunately, some researchers who conduct such studies tend not to publish their
results if they show no sex difference.)

What about humans? And what about brain differences due
to sexual orientation, as well as sex? A group led by Simon
LeVay reported in 1991 that a tiny nucleus of cells in the hypothalmus was different in size ("dimorphic") based on both sex
and sexual orientation.[20] This area of the brain, known as the
"third anterior nucleus of the anterior hypothalmus" (INAH 3),
is smaller than the dot at the end of this sentence. In fact, INAH
3 is not so much a well-defined brain structure as a grouping
(or nucleus) of tiny discrete neuronal cells.

LeVay's study has received a great deal of media attention
because it reported that INAH 3 tended to be significantly
smaller, on the average, in gay men than in heterosexuals—
providing still more evidence for a human brain difference
based on sexual orientation. Furthermore, LeVay reported that

24   INAH 3 in gay men tended to be about the same size as in heterosexual women, agreeing with the earlier idea that gay men have "feminized" brains. As can be seen in figure 2.3, it is only for INAH 3 that these differences are found; no significant differences are seen in the other three INAH regions.

Three other observations stand out when we examine LeVay's data for INAH 3. First, the data set is very small—the dots represent the measurements for individual brains. Second, the distribution in size for INAH 3 overlaps between homosexual males (HM) and the heterosexual males (M). Thus, even though the average size is different between these two groups, there are three cases for each group that fall in the typical size range found for the other sexual orientation. Third, the average size of INAH 3 based on the dot distributions in figure 2.3 is between two and three times larger in heterosexual men than in gay men. According to LeVay, the probability of a difference this large arising by chance is no more than one in a thousand.

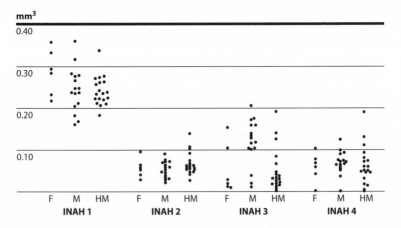

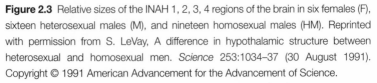

Figure 2.3 Relative sizes of the INAH 1, 2, 3, 4 regions of the brain in six females (F), sixteen heterosexual males (M), and nineteen homosexual males (HM). Reprinted with permission from S. LeVay, A difference in hypothalamic structure between heterosexual and homosexual men. *Science* 253:1034–37 (30 August 1991). Copyright © 1991 American Advancement for the Advancement of Science.

25   The issue of statistical significance is an especially important one any time we spot a difference between two sets of data whose magnitude was not previously predicted. Many anomalies can crop up by chance, and it is very difficult to evaluate the statistical significance of an unpredicted anomaly *after* it has been spotted. The only reliable test of statistical significance for a strictly empirical observation is whether follow-up observations show the same effect and to the same degree.

One brain researcher, William Byne,[21] who had been a vocal skeptic of LeVay's work, has recently confirmed that a male—female INAH 3 size difference does exist. But, Byne's attempt to replicate LeVay's finding for sexual orientation is problematic. Byne compared a sample of brains from heterosexual and homosexual men and found that the average INAH 3 volumes (divided by total brain weight) was 26 ± 10 percent greater in heterosexuals. This result is statistically significant at the level $p < 0.05$ (one chance in 20).

On the other hand, Byne's 26 percent difference between the INAH 3 volumes in heterosexual and gay men is far short of the factor of two or three that LeVay found, and so it is only a partial confirmation. Also, Byne, unlike LeVay, compared the INAH 3 regions by counting individual neuronal cells in them. On the basis of cell counts rather than overall volumes, he found no significant differences between gay and heterosexual men. Thus, at best, Byne's work can be said to offer only a weak partial confirmation of what LeVay earlier reported.

A serious problem with both studies is that these kinds of brain measurements cannot yet be made on living people. Most of the gay men whose brains were measured in both studies died of AIDS. Perhaps LeVay was merely observing an effect of AIDS on the size of INAH 3. However, neither Byne nor LeVay found a correlation between the HIV status of the deceased and the size of their INAH 3 regions. In conclusion, the evidence for brain structural differences based on sexual orientation remains tentative.

Whatever future research should reveal about the reality of brain structural differences based on sexual orientation, there is

26   one overriding question: do these differences say anything about the *causes* of homosexuality? At least two possible answers suggest themselves, corresponding to opposite directions of causation. One possibility is that individual men who are born with smaller INAH 3 regions are more likely to be gay as a result of that fact, which would correspond to what is found in rodent studies. A second possibility is that the size of the INAH 3 region is initially undifferentiated by sexual orientation, but homosexual experiences or fantasies somehow make this region of the brain smaller in gay men. This latter scenario is not as crazy as it sounds, since it is well known that the structure and functioning of parts of our brain change in response to experiences and learned behavior. Unlike our brains, which change during the course of our lives, our genes are part of our original equipment. So, perhaps the best way to establish whether homosexuality is innate is to study our genes.

Search for a "Gay Gene"

A promising place to look for evidence that homosexuality may be genetic is research on identical twins. Every study of male twins since the early 1950s has shown some degree of heritability for homosexuality, with the results for women being more ambiguous. One large study by Bailey and Pillard reported that just over 50 percent of the time when one twin brother was gay, so was the other.[22] A "glass half-full" interpretation of this result might stress that these gay co-twins were therefore almost 10 times more likely to be gay than the general population. The "glass half-empty" interpretation of this study could note that it showed that at least 50 percent of the time environmental factors rather than genetics must have been at work.

The true figure for environmental causation could even be larger than 50 percent, because usually identical twins are raised in very similar environments, and a high degree of agreement in their sexual orientation might be expected in part on that basis. (Some studies have been done on the sexual orienta-

27   tion of twins reared apart, but the small number of cases where one twin is gay makes it difficult to draw any firm conclusions.) One problem with twin studies of sexual orientation is selection bias, particularly if the study's subjects are volunteers. The Bailey and Pillard study got its subjects by advertising in gay-oriented magazines. We might, therefore, suspect that twins where both are gay would be more likely to volunteer for the study than if only one twin is gay. Bailey and Martin did another study in 1995, which relied on an Australian twin registry to avoid the selection bias problem.[23] Sure enough, they found a much smaller "concordance rate" for male identical twins—only 20 percent of male gay co-twins were themselves gay, with similar rates for females—showing that selection bias is indeed a problem in twin studies that rely on volunteers.

Despite such problems, the overall results of twin studies suggest that genes play a significant role in determining sexual orientation. This possibility gains more credibility by the finding that gayness seems to have a maternal link in family trees. Such a maternal linkage could mean that homosexuality is a trait inherited from the mother, just like the most common type of color blindness and male baldness. Dean Hamer and his colleagues looked at the family backgrounds of 76 gay men, a study described in Hamer's excellent book *The Science of Desire.*[24] They found that only male relatives on the mother's side of the family had any higher than chance levels of being themselves gay.

Maternal inheritance implies that the gene for the trait must be on the female (X) chromosome, since males inherit that one only from their mother. Narrowing the location of a possible gay gene to one of 24 chromosomes is important, but it is a very long way from locating the gene itself, since there are thousands of genes on that chromosome. In fact, all the maternal linkage data really does is to suggest that the X chromosome is a promising place to look for a gay gene, *if* one exists.

Hamer's group proceeded to conduct their search by a method known as "linkage" analysis. The idea goes something like this. When the sperm and egg unite to form an embryo, a

28   new set of 24 chromosomes are formed by combining those from each parent—except for chromosome X, the "female" chromosome. Since males have one X and one Y chromosome, and females have two X's, all males inherit their X chromosome from a scrambled ("recombined") version of the two versions that their mother has. This recombination process results in the new chromosome, having alternating regions that originated from the mother's two versions (A and B) of the X chromosome (see figure 2.4).

On the chromosome biologists locate a series of "markers" (identifiable regions that show a significant variability from person to person.) These markers are likely to be different on each version of the X chromosome the mother possesses. Thus, if a particular gene lies on the chromosome near one of these markers, there is an excellent chance that during the recombination process if the marker winds up being part of the newly formed chromosome, so would the gene. Putting it differently, if we find two sons who have the same version of a particular marker, there is an excellent chance they also have the same versions of any genes that lie near that marker on the chromosome.

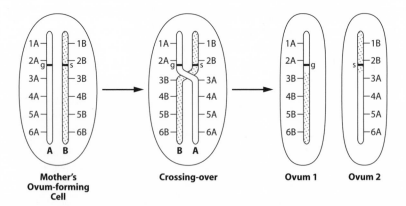

Figure 2.4 Two new chromosomes formed from the recombination of the two versions of the mother's X chromosome. Reprinted with permission from S. LeVay, *Queer Science: The Use and Abuse of Research into Homosexuality.* Cambridge, MA: MIT Press, 1996. Copyright © 1996 MIT Press.

29   If biologists suspect that a particular trait has a genetic basis, they do a "linkage analysis." They locate pairs of siblings that share this trait (and presumably also share the gene for it), and then see how often they share particular markers along the chromosome. If the sharing occurs no more often than chance, then either the particular marker is not near the hypothetical gene, or else the gene doesn't really exist. In this context, "chance" means 50 percent, because any particular marker is equally likely to have come from each of the mother's two X chromosomes, which very likely have different versions of the marker.

Hamer's group did their 1993 linkage study on 40 pairs of gay brothers. Essentially, they asked how many of the brothers shared each of 22 markers along the X chromosome. In most cases the sharing was at no more than chance level, with half the brother pairs sharing the marker. However, for the five markers in the region called Xq28, located near the end of the chromosome, the incidence of sharing was well above 50 percent.[25] In fact, Hamer's group found that the last five markers were shared by 33 out of 40 of the brother pairs—meaning 33/40 = 82 percent sharing. If this result were simply due to chance, it would be equivalent to getting 33 heads after 40 coin tosses, and so chance seems an unlikely explanation.

Hamer's group also looked at lesbian sisters of the gay brothers, and found no markers on the X chromosome that had any significant sharing above chance levels.[26] Hamer claims that the high degree of sharing found for the gay brothers would happen by chance no more than once in 100,000 times. This figure represents the chances of getting at least 33 heads (or tails) in 40 coin tosses. This one in 100,000 chance, however, is clearly an overestimate of the statistical significance of the result, because the anomaly could have been found on any of the 22 markers that were looked at on the X chromosome—or for that matter any of around 500 regions on the full genome.

A more reliable estimate of the probability of finding the observed degree of sharing anywhere on the genome by chance

30   is, therefore, 500 times 1/100,000 = 0.005. Another way to say the same thing is that if the observation was just a random fluke, it would not occur 99.5 percent of the time we looked.

A Digression: The Face on Mars and the Case of the Missing Dots

There are many ways to illustrate the tendency of the human mind to find patterns where none exist. One example would be the "face on Mars" seen in some early NASA photographs of the planet's surface (figure 2.5, left). Some people believe that the "face" was constructed by Martians, and hence is evidence for intelligent life on Mars. However, higher resolution images of the "face" area of Mars clearly show that it doesn't look much like a face anymore (figure 2.5, right). (For a contrary opinion, including a large set of NASA photographs of the Martian surface claimed to provide other evidence of artificial structures constructed by Martians, see www.metaresearch.org/default.asp, a web site maintained by Tom VanFlandern.) It's rather difficult to estimate the chances of seeing a face

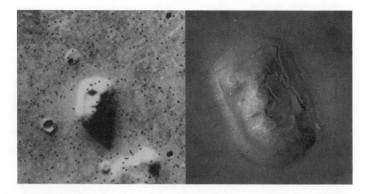

Figure 2.5 The "face" on Mars, as originally seen in a 1976 photo (left), and as seen in a subsequent higher resolution photo in 2001 (right), courtesy of NASA/JPL/Malin Space Science Systems. Part of the reason for the disappearance of the "face" is the change in the angle of the sun illuminating the feature.

31 when looking at a random Martian landscape or a cloud, because our minds are programmed to see such patterns.

Here we'll take a look in detail at another example of a tendency to see patterns when none exists involving a random distribution of dots. Figure 2.6 shows a pattern of 100 dots generated using a random number generator. The distribution of dots is uniform, but it is subject to the expected statistical fluctuations. If you weren't aware of the tendency of random dots to cluster *somewhere* you could easily be misled into seeing meaningful patterns when none exist.

Let's say the dots represent measurements of some kind, and you have a theory predicting they will be absent from some rectangular region, but you don't know its location. So, you look at the pattern to see whether any rectangular hole appears. The small rectangle, which occupies 8 percent of the total area, shows the largest dot-free region in the figure. How unlikely is this absence of dots? You might reason that since the rectangle occupies 8 percent of the total area, it should contain 8 dots—8 percent of 100. Finding zero dots when you expect eight is an extremely unlikely occurrence.

Here's how you might try to calculate the chances. The probability of any one out of the total 100 dots appearing outside the rectangle is 0.92, since 92 percent of the total area lies outside of it. Therefore, the probability that every one of the 100 dots will lie outside the rectangle is

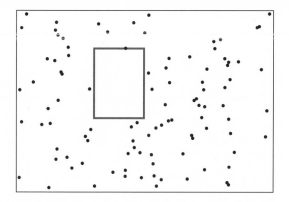

Figure 2.6 One hundred dots produced by a random number generator. The small rectangular region outlined in the figure shows the largest dot-free region found in the figure.

32 $p = 0.92^{100} = 0.00024$, or only one chance in 4180. On this basis you might conclude that your theory predicting an area devoid of dots has been confirmed. But you would be sadly mistaken!

Do you see the fallacy in the preceding argument? It must be non-sense, because there was nothing special about this particular random number pattern—all such patterns contain both clusters and vacant regions. We have made the error of an "informed choice"—picking a rectangular region of the plot to examine *after* observing the pattern. The previous statement about the remote chance of finding zero dots when eight were expected would be true only if you had identified the particular rectangle you wanted to look at *before* you saw the dot plot. Moreover, it is difficult to know what factor to multiply the previous value of p by to find the true probability here, because we chose the length, width, and position of the rectangles after seeing how the points clustered and where vacancies appeared. The extreme bias that we used in selecting this particular rectangle can be easily demonstrated by adjusting its boundaries outward very slightly to include some of the surrounding dots.

Let's carry our analysis a further step. Suppose once you have found the absence of points in the rectangular hole (which you might mistakenly believe to be real and not random), you then carry out another "confirmation" experiment and get a new dot distribution. Suppose the new experiment winds up giving eight dots in the rectangle—exactly what chance would predict, since the patterns really are assumed to be random here.

The proper conclusion is that no absence of dots has been observed in the rectangle, based on the results of the replication experiment. However, if you were to combine the replication experiment with the original one, you would obtain a total of eight dots in the rectangle, compared to the expected number of $2 \times 8 = 16$. Finding as few as 8 dots when you expected 16 is still quite unlikely, but it is again a *bogus* result, because we cannot combine the original and replication experiment. In summary, we can draw three lessons from this "case of the missing dots":

1. Making informed choices can greatly inflate the statistical significance of an observed anomaly.
2. Choosing the correct number to multiply by to find the true probability (that takes informed choice into account) can be tricky.

33   **3.** When informed choices are made in the original experiment, statistical tests that combine replication experiments with the original one will give an inflated overall statistical significance.

Does a Gay Gene Really Exist?

The best way to evaluate whether Hamer's gay gene is genuine would be to see whether follow-up studies confirm it. Three follow-up studies have been done at the time of writing, one by Hamer's group and two by independent groups.[27] The overall results are underwhelming. Two studies showed some excess sharing of Xq28 markers above the chance level (50 percent), but not nearly the 82 percent cited by Hamer's original study, while the third study showed *below* a chance level of sharing. Combining the data from the three follow-up experiments, we find the fraction of gay brothers sharing their Xq28 markers is $81/138 = 59$ percent. Given the small number of brothers in the three studies this percentage does not differ from the chance level of 50 percent by a statistically significant amount. On the other hand, it is very significantly below Hamer's original result (82 percent). Although the combined results of the three follow-up studies offer very little confirmation of Hamer's Xq28 hypothesis, Hamer is not ready to throw in the towel. In support of Xq28, he makes the following arguments:

1. Of the three follow-up studies, only the Rice study really conflicts with the original study, since the sharing values found by the other two were above the chance level (50 percent) by a statistically significant amount ($p < 0.04$).
2. The Rice study, which gave a negative result, is flawed because it did not systematically exclude families showing evidence for possible nonmaternal inheritance (e.g., families with gay fathers).
3. The Hu study (also done by Hamer's group) showed additional confirmation for the Xq28 hypothesis, because it

34   also looked at heterosexual brothers, and found they had less than 50 percent sharing of Xq28 markers with the gay brothers, as would be expected.

4. The most objective way to assess the overall evidence is to look at *all* the studies (including Hamer's original study along with the three follow-ups), and this would show a statistically significant combined result.

Each of Hamer's points is debatable, particularly his last one. As we have shown in the example in the digression about seeing patterns when none exists, if we combine replication experiments with an original study in which an "informed choice" is made, we incorrectly inflate the overall statistical significance. What of Hamer's second rebuttal point, namely that Rice's study was biased because it did not exclude nonmaternally linked gay relatives? Needless to say, Rice and his colleagues dispute Hamer's claim that their study was biased, and they note that their negative result would change little had they dropped the two families with gay fathers. Finally, Rice and his colleagues note that the whole idea of a maternal linkage of a "gay gene" on the X chromosome is open to question, because a new family pedigree study done since Hamer's original study does not show any maternal linkage for gayness among relatives.[28] And, they note that the very idea of a gay gene remaining in the gene pool after many generations seems highly implausible.

How Might a Gay Gene Remain in the Gene Pool?

As noted earlier, this last criticism of the gay gene idea is the reason many lay observers might think it justifies the "crazy idea" label. How could a gay gene not disappear from the gene pool, given that gay men have far fewer children than heterosexuals? Several answers have been suggested to this puzzle. One speculative possibility is that even though gays have few children, they can pass their genes on to future generations by helping close relatives raise their children and make it more likely that those children survive and reproduce.

35   This idea is an illustration of W. D. Hamilton's argument that evolution doesn't reward individual organisms that reproduce most successfully; rather it rewards the *genes* that cause organisms to behave in such a way that the genes make it into the next generation. Thus, for example, a gene for altruistic behavior (among close relatives) could become prevalent in the gene pool, provided it had sufficient benefit to all its carriers collectively. (J.B.S. Haldane, the biologist, when asked if he'd give up his life for his brother, is said to have thought for a moment, and after doing a brief calculation, said that no, but he would for two brothers or eight cousins, since in that case the same fraction of his genes would survive.)

The idea that genes for altruistic behavior can thrive in the gene pool seems quite plausible, at least for close relatives, but its applicability here to a gay gene seems very questionable. The problem is that since gays have far fewer children than heterosexuals, their help in raising children of close relatives would need to go way beyond mere occasional assistance to make the difference in the survival of their genes. For example, your neices and nephews have 25 percent of their genes in common with you. Therefore, to pass on the same fraction of your genes to future generations, for each son or daughter that you didn't have, you would need to keep four nieces or nephews alive to reproductive age. Those four nieces or nephews would have had to die without your assistance.

An alternative theory as to how a possible gay gene might have managed to maintain itself in the gene pool seems more plausible. In the alternative theory, a hypothetical gay gene, "G," has a different manifestation in male carriers and female carriers. Suppose that in males gene G causes the carrier to be attracted to men, while in females G causes them to be even more attracted to men than is usual for women, making the women "*hyper*heterosexual"—and hence more likely than usual to reproduce. A hypothetical lesbian gene, "L," could conceivably work in a similar manner, making both sexes more than usually interested in having sex with women (though you might believe that's impossible for most males!)

36 Still another possibility is that the version of the gene causing homosexuality is recessive, similar to that which causes red hair. This means that it fails to cause homosexuality and is "invisible" to the process of natural selection unless it is paired with a second gay version of the gene. If the recessive gene were to convey some unrelated evolutionary advantage of greater positive benefit, it would remain in the gene pool.

One example of this idea concerns the serious genetic disorder afflicting African Americans known as sickle cell anemia. Red blood cells of the sickle variety are misshapen (like sickles) and cannot carry the normal amount of oxygen. The sickle version of the gene is recessive, so if you only get one of them (from one parent), you are merely a carrier, but do not have the disease. However, such carriers do enjoy the benefit of some extra resistance to malaria, a completely unrelated condition. The gene survives in the affected population, because its benefit outweighs the harm, in terms of allowing people to reproduce. This example certainly shouldn't imply that homosexuality is a some kind of disease, like sickle cell anemia, that needs some positive benefit to offset it. Rather, the point is that being gay does imply a lower-than-average level of reproductive success. Without some compensating factor that improves the chances of one's genes of making it into the next generation, any gay gene would disappear from the gene pool over time.

The three preceding explanations of how a gay gene might remain in the gene pool are speculative, and the issue might be best addressed more fruitfully *if* a gay gene is ever found. So far, all that has been demonstrated is that for a certain class of gays (in families with two gay brothers), there is some possibility that Xq28 may be a region of the X chromosome on which a gay gene could conceivably exist. Overall, I think it fair to say that Hamer's hypothesis of a gay gene located somewhere on Xq28 is still alive, but hanging by a thread. Even if the Xq28 results should be confirmed, Hamer notes that a gene in this region would probably account for no more than perhaps 10 percent of homosexuality in males, with much of the remainder determined by additional genes at other locations. Clearly, what is

37 needed to settle the question would be a study with greater statistical significance than those done so far, searches that include chromosomes other than X, and searches at the level of individual genes—all of which Hamer's group is currently conducting.

Although human sexual orientation might be too complex to be based on a single gene, that may not be the case for simpler organisms such as fruit flies, where there is the added advantage of being able to make experimental genetic modifications. One might suppose that fruit flies have little in common with humans, but scientists believe that two-thirds of their genes have human counterparts. Recently, fruit flies have been genetically modified so as to switch from heterosexual to homosexual behavior by altering a single gene.[29] Strangely, the flies' sexual behavior is linked to environmental temperature. Above 30°C modified male fruit flies lose interest in mating with females and become receptive to other males. Once the temperature reverts back to below 50°C, the flies' sexual preference also changes back to heterosexual.

To give a credibility rating to the idea that homosexuality is largely innate, we need to consider the strength of all the various kinds of evidence discussed earlier, which is summarized in table 2.1. Some of the strongest evidence comes from animal

Table 2.1

Summary of the Evidence for Homosexuality Being Largely Innate

Type of Evidence	Strength of Evidence
Recollections of gays	Weak (subjective)
Connection with handedness	Strong, but circumstantial
Early age at which behavior appears	Moderate
Adopted children of gay parents	Moderate
Identical twin studies	Strong
CAH girls (prenatal hormones)	Strong
Brain anatomy (INAH 3)	Weak (replicable?)
Genetic linkage analysis (Xq28)	Weak (replicable?)
Rodent studies	Strong (human applicability?)
Fruit flies	Strong (human applicability?)

38   studies, whose applicability to complex human behavior could be questioned. But, looking at all the evidence collectively, there seems to be a reasonable likelihood that homosexuality is innate to a significant extent in humans as well. Therefore, my rating for the idea that homosexuality is primarily innate is zero flakes—meaning not flaky at all.

What Are the Implications?

The percentage of Americans accepting the idea that homosexuality is something a person is born with has increased dramatically since Hamer's research on the gay gene—from 17 percent in 1982 to 40 percent in 2001, according to Gallup Polls (figure 2.7).[30] If it should be confirmed that there is a gay gene (or more likely a combination of genes), conventional wisdom says that would be good news for gays, who it would be assumed could no longer be accused of choosing to adopt an immoral lifestyle.

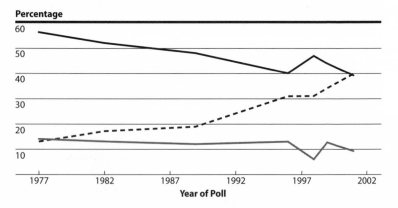

Figure 2.7 Cases of homosexuality. Gallup Poll results on the percentages of Americans who believe that homosexuality is due to upbringing or environment (upper solid curve), something a person is born with (middle dashed curve), or both factors (lower light curve). The percentages for a given year don't add to 100% because the "neither" and "no opinion" responses have not been included here.

39   But that assumption misreads the nature of morality, the politics of gay and antigay forces, and the meaning of a gay gene.

If there is a gay gene, it would not necessarily cause everyone who has it to become gay; rather it would probably only make it more likely. To make moral choices in life, we sometimes need to defy temptation. Antigay conservatives confronted with a gay gene would likely stress the importance of those with the gene being even more vigilant not to give in to their temptation. Thus, even though the percentage of Americans who believe that homosexuality is something a person is born with has dramatically increased during the last two decades, the percentage who believe that homosexual relations between consenting adults should be legal has changed much less since 1989 (figure 2.8).[31]

If homosexuality is genetic, increased levels of public acceptance could paradoxically lead to a decrease in the fraction of gays in the population over time. For example, in times gone by the greater stigma against gays made it more likely that gay men and women would remain in the closet and not acknowledge their sexual orientation, even to themselves, until after a

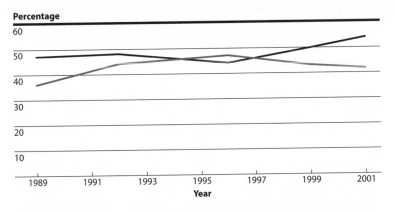

Figure 2.8 Should gay sex be legal? Gallup Poll data on responses to the question of whether homosexual relations between consenting adults should be legal. Dark curve is the percentage of respondents answering "yes," and light curve is the percentage answering "no."

40 failed marriage. Since many failed marriages still may have resulted in children, they represented one vehicle for a possible gay gene to propagate into future generations. Conversely, if the acceptance of gays adopting children should increase, these adopted children would not carry a gay gene with any greater frequency than the general population.

The politics of a gay gene could also make for some very strange "bedfellows." Just think of a pro-life and antigay woman who learns to her horror that her fetus has the gay gene. And what about a pro-choice but pro-gay woman who has dreams of someday becoming a grandparent. Would these women's attitudes toward abortion trump their attitudes toward homosexuality? Currently, abortion for the purpose of selecting the sex of a baby is quite common in some countries, particularly China and India. In China forced abortions and female infanticide have been used to limit family size to one child.[32] It is difficult to imagine that abortion on the grounds of sexual-orientation selection would be less common than abortion based on the sex of the fetus. If such abortions were to become routine, homosexuals could, over the course of a few generations, find themselves a steadily declining fraction of the population.

3 Is Intelligent Design a Scientific

Alternative to Evolution?

NORMALLY, you might be content to leave the question of whether a theory is scientific up to scientists. Such may not be the case, however, for theories such as evolution and intelligent design, which relate to deeply held world views and religious beliefs. A surprisingly large gap exists between scientists and the American public regarding their acceptance of the theory of evolution according to 1997 polling data (Table 3.1).[1] A similar gap exists regarding belief in God—40 percent among scientists versus 90 percent among the public generally. In contrast to most scientists, a large majority of the U.S. public believes either that evolution did not occur or that it had to be directed or guided by God ("theistic evolution"), with a significant variation depending on a person's level of education. However, even among those with college degrees, only one in six accepts "naturalistic evolution" (evolution with no help from God). In fact, when compared to citizens of 16 other industrialized "Christian" countries in a 1991 survey, Americans were found to have the smallest percentage of people who believed that humans evolved from earlier species of animals, and the highest percentage of people believing in miracles.[2]

Table 3.1
Percentage of Different Groups Believing in Various Theories

Group	Biblical Creation	Theistic Evolution	Naturalistic Evolution
College grads	25	54	16.5
No high school degree	65	23	4.6
All U.S. public	47	40	9
Scientists	5	40	55

42   There are two views of the nature of the gulf between scientists and the public, depending on which side of the debate you're on. In the view of many academic scientists, the problem is strictly one of inadequate public understanding of evolution, a theory that is a cornerstone of modern biology, and which has been well tested, although not always well taught. In this view, the idea of "intelligent design" (God-guided evolution) is merely an updated version of creationism—a viewpoint that seeks to promote traditional Christian religious teaching under the guise of science. Many scientists also believe that intelligent design is no more testable than old-fashioned Biblical creationism, and is a matter of faith rather than science.

Conversely, many of those in the intelligent design (ID) movement make precisely the same claims about Darwinian evolution, namely that it, rather than ID, is simply a point of view that cannot be tested scientifically. Further, they claim that Darwinian evolution is an attempt to promote a materialistic or naturalistic philosophy that leaves no room for God. Cognizant of the low level of belief in God on the part of elite scientists, the ID movement considers the scientists' support for evolution to be merely a reflection of their materialistic worldview.

To decide whether the debate between naturalistic and God-guided evolution is a matter of science or world view it is helpful to consider the roots of the present intelligent design movement. These roots are largely a reaction to Darwin's theory of evolution by fundamentalist branches of Christianity and Islam.[3] Interestingly, Charles Darwin himself was a devout Christian who set out on his voyage on the *Beagle* to find support for the biblical account of creation, and he later tried to assure readers that his theory of evolution need not be in conflict with religion.

Darwin's Theory of Evolution in a Nutshell

Charles Darwin published his theory of evolution in his book *The Origin of Species* in 1859. The theory, which was independently proposed by Alfred Russell Wallace, essentially says that all life has a common ances-

43 try and that it evolved through the process known as natural selection, which is often summarized by the phrase "survival of the fittest." Unfortunately, that well-known phrase is widely misunderstood and misrepresented. That's because fitness here does not refer to specific traits such as strength, speed, and cunning, but only the ability to survive and produce offspring. Thus, species whose members are killed off in large numbers by predators can still be regarded as being fit if they produce an even larger number of offspring.

Natural selection is the process in which traits favorable to survival get passed down to future generations simply because organisms with those traits are more likely to survive and reproduce. Hence, the process is completely automatic. Since traits are now known to be inherited through genes, the idea of passing on favorable traits has been labeled the "selfish gene."[4] The idea is that we as individual organisms are in a sense merely receptacles for the genes responsible for those favorable traits. Thus, while the phrase "survival of the fittest" conveys an impression of extreme selfishness, natural selection can in fact account for the evolution of altruistic behavior among related individuals. This seems paradoxical, because when individuals behave altruistically, they may place their own survival in jeopardy. But, genes for such behavior can get passed down to future generations if an act of altruism results in the survival of a relative, especially an offspring.

The process of evolution through natural selection is "blind" or without purpose, because it happens entirely automatically. Thus, for example, there is nothing in evolution that pushes a species toward the development of greater complexity. It could happen that a species becomes more complex over time—say by developing a greater brain capacity—but that development would occur only if it promoted greater reproductive success in individuals. Those who subscribe to the theory of evolution believe that it can account for the entire development of life, including both changes within species and the development of new species. Some evolutionists also suspect that the theory can even account for the origin of life from nonliving organic molecules.

Although Darwin's theory of evolution was controversial when it came out in 1859, many Christian fundamentalists ini-

44   tially did not see it as a grave threat. The controversy between religion and evolution didn't really heat up in the United States until 1925, with the efforts of three-time presidential candidate William Jennings Bryan. Bryan was also Secretary of State in the Wilson Administration, and he was a moral and religious crusader with formidable oratorical skills. He was one of the key protagonists in the Scopes "Monkey Trial" in which John Scopes, a high school science teacher, challenged the Tennessee law forbidding the teaching of "any theory which denies the story of the Divine creation of man as taught in the Bible."[5] Other court cases followed in the years after Scopes. As part of their legal strategy, creationists, led by professor Phillip Johnson of the University of California Law School (Berkeley), have tried to frame the debate over evolution by portraying intelligent design as a genuine scientific alternative to evolution, as described in Johnson's 1991 book, *Darwin on Trial.*

There is no intrinsic conflict between evolution and religion, and many scientists who completely accept the theory of evolution are also devout believers in traditional forms of religion—though not in a literal interpretation of the Bible. The idea that blind evolution leaves room for, at most, a hands-off God who merely started the whole show with a Big Bang may seem plausible to some evolutionists, but it is beautifully disputed in Kenneth Miller's 1999 book, *Finding Darwin's God.*[6] Although it may sound paradoxical to some readers, Miller, a Brown University biologist and a devout Catholic, finds in Darwinian evolution strong support for a God that cares for us and is intimately involved in our lives. Thus, Miller's view demonstrates that evolution and traditional religion need not be in conflict. Here are four possible viewpoints on a spectrum of beliefs regarding evolution:

1. *Naturalistic evolution.* Life evolved strictly through purposeless natural selection.
2. *Theistic evolution.* God might be operating behind the scenes, but all that we actually observe in the evolution of life is consistent with blind natural selection.

45   3. *Intelligent design.* All life descended from a common ancestor, but the mechanism of its evolution could be some
combination of natural selection and intelligent design,
with evidence for the latter being observable.

4. *Young-Earth creationism.* The Earth is less than 10,000
years old as inferred from the Bible. Scientific evidence
that appears to contradict this fact, such as the fossil
record, radioactivity, and the radiation left over from the
Big Bang, were created by God either to test our faith or
for purposes we cannot fathom.[7]

It would be a mistake to assume that people who subscribe to
creationist beliefs are simply a bunch of ignoramuses. As geo-
scientist Steven Dutch has noted, many creationists

inhabit a parallel universe to that of science with its own
bodies of evidence, observational data, and technical literature. Communication between parallel universes is possible,
but only if there is a portal between them. For some believers, the portal is only through the Bible and effective communication may not be possible. . . . To believers in [the cre-
ationist] milieu, the average evolutionist scientist looks like
a flat-earth believer, someone wholly unacquainted with the
technical literature, yet arrogantly demanding that everyone
else discard well-established ideas to accept a new theory. To
be as blunt as possible, they see us as the crackpots."[8]

We can simplify the debate over evolution by choosing to
place a line between the naturalistic and theistic evolutionists
on one side and the intelligent designers and creationists on the
other. The basis of that arbitrary dividing line is that the first
group believes that intelligent design is unobservable in nature,
and therefore outside the realm of science, while the second
group would dispute this claim. (Thus, that dividing line also
reflects the basic question asked in the title of this chapter.)

Whether or not intelligent design is a scientific matter may
strike you as strictly a question of no real-world consequence.
After all, why should you care if many scientists regard ID

46 merely as a creationist wolf in sheep's clothing, while intelligent design folks regard it as real science? One reason you *should* care is a 1987 Supreme Court ruling against the *Creationism Act* passed by Louisiana. This state law required public schools to teach biblical creationism along with evolution. In overturning the *Creationism Act*, the Court noted that the law would impermissibly introduce religion into the public schools. The Court ruling, however, also noted that states might still be free to mandate the teaching of *other* scientific theories as alternatives to evolution. Thus, if ID theory really is science, whether true or false, it would, according to the Court, deserve its rightful place in the curriculum alongside the theory of evolution, something favored by a large percentage (68 percent) of Americans.

What Does ID Have to Say about God?

Some ID adherents make no claim as to the identity of the "designer," even suggesting various other possibilities besides God, such as space aliens and time travelers! This disinclination to name God as the designer might be expected if they wish to make a case that ID is science, not religion. Such agnosticism aside, we can ask whether the ID debate really is about the nature of God.

Evolution by natural selection might suggest a randomness to the evolutionary process that belies a purposeful God. Given all the unpredictable contingencies, evolution would not have led to humans if the tape were replayed—a point made forcefully by Stephen Jay Gould. Who knows?—If a large asteroid had not struck the Earth 65 million years ago, intelligent dinosaurs, not humans, might be the planet's dominant species. If we are the apple of God's eye, as traditional religions suggest, on what basis could we expect that anything vaguely resembling humanity would have evolved, given blind evolution and random environmental changes?

As Kenneth Miller argues, the seeming randomness of evolution need not be incompatible with a purposeful God. Clearly,

47   an omnipotent God could work His will by performing miracles and contravening natural laws. But Miller notes that there is no reason to believe that an all-powerful God could not also work His will while remaining within the framework of natural law. Miller and others have cited the unpredictable processes of quantum physics as one way that God could achieve His purposes without being detected. He further notes that if you happen to believe God is influencing the events in the world today without contravening natural laws, why would He need to contravene natural laws for the past evolution of life? If God is omnipotent, He surely could have worked His will within a strictly Darwinian scheme without leaving any intelligent design clues.

Still, Darwinian evolution strikes some people as imputing to God a number of unattractive features, including wastefulness and cruelty. Doesn't it seem wasteful to bother creating one species after another, only to have the vast majority of them become extinct? And, if we are the main point of the show, why take as long as 15 billion years after creating the universe to get around to creating humanity? On the cruelty score, what kind of God would use blind evolution based on a survival of the fittest scheme, with all the suffering that often entails for the unfit? On the other hand, these sorts of questions about the nature of God raise no deeper puzzles than other questions that are independent of evolution: Why does God allow evil in the world? And, if God created the world, where did God come from? For some of us these questions are so puzzling that the very existence of a personal God may be an unresolvable dilemma. As fascinating as these questions are, however, we shall put them aside to return to our main question: is intelligent design a scientific matter?

Are Some of Life's Structures "Irreducibly Complex"?

One concept devised to show evidence for intelligent design in nature has been the idea of irreducible complexity. This term

48 was coined by Michael Behe, a professor of biochemistry at Lehigh University and author of the 1996 book *Darwin's Black Box*.[9] According to Behe, an "irreducibly" complex system is one that would cease to function if any one of its components were removed or even slightly altered. Behe maintains that certain biomolecular structures and functions are irreducibly complex, and that, as a result, their evolution by natural selection would have been exceedingly improbable, even given enormous spans of time. Three examples of such structures that Behe cites are the cell, the flagellum that bacteria use to propel themselves, and the complex mechanism for clotting blood.

Critics of intelligent design note that Behe's idea is simply a dressed-up version of the argument advanced by theologian William Paley (1743–1805), whose favorite example of a clearly designed object was that of a watch found in a field. According to Paley, anyone finding a watch would know that it was a designed object and not a natural one. Supposedly, even an alien unaware of its function could infer that the watch had a designer. (However, I do wonder what highly intelligent "dolphin-like" aliens who never built anything or aliens who always used a technology relying on irregularly shaped components might make of a watch [figure 3.1].) Darwin was aware of Paley's argument when he came up with his theory of evolution, and found it no more persuasive than critics of design find it today.

Figure 3.1 On a planet where timepieces were designed by Salvador Dali, an ordinary earthly watch might go unrecognized as a designed object. Image of "Soft Watch at Moment of First Explosion," by Salvador Dali. Copyright © 2002 DEMART PRO ARTE B.V.

49   For example, consider a complex structure such as the eye, which requires a number of closely matched components to function: pupil, lens, retina, etc. Darwin understood that even though a fully evolved eye required all these components, the organ would have some degree of functionality if one of them, say the lens, were missing. In fact, some primitive organisms such as algae, find it very useful to be able to detect light merely with the aid of a light-sensitive spot. Such primitive limited-functioning precursors to the eye told Darwin that eyes could evolve. In fact, in his *Origin of Species*, Darwin points to a series of increasingly capable eyes in animals to argue his case for the evolution of this organ. Oxford University biologist Richard Dawkins has also taken up the challenge and shown how the eye could have formed during the course of evolution by providing a step-by-step possible sequence.[10]

Intelligent designer Behe, however, argues that his updated concept of irreducible complexity is not something Darwin could have brushed aside so easily. In the first place, Behe is concerned with structures at the molecular level, of which Darwin was unaware, not large-scale features like eyes. More importantly, Behe argues that the structures he cites really are irreducibly complex, unlike eyes, which admittedly may have a limited degree of function with one component missing.

Behe's claim that such structures could not have evolved is based on the assumption that evolution requires a series of small incremental steps based on random mutations, each of which makes the organism more fit to survive. If a structure has many closely matched components, presumably the independent components would all need to mutate separately in specific directions and to the same degree, otherwise they would not remain well matched. The likelihood of many separate independent mutations occurring at the same time in the right way is vanishingly small, argues Behe.

One scientist who has disputed Behe's claim is Brown University biology professor Kenneth Miller. He examines two of Behe's examples of irreducible complexity. Regarding the flagellum—the "outboard motor" of bacteria—Miller notes that

50   there are many simpler predecessors that indicate an evolutionary pathway. But Behe notes that is not the same as spelling out
a specific pathway consistent with natural selection operating
at each step.

The second example Miller challenges is that of blood clotting. Blood clotting in mammals takes place in a cascade of
steps, each one triggering the next, to respond to a small initial
cause—the detection of a leak in the system. But the system
needs to work perfectly, because if it is triggered in the event of
a false alarm, blood stops flowing when it is needed! Simpler
clotting schemes exist in other organisms, such as lobsters,
which don't require a cascade of steps, and Miller argues that
the cascade steps could have been added by evolution
through a process known as gene duplication. But Behe claims
that the specific intermediate steps are suggested only in a
hand-waving manner that does not explain how each intermediate step would lead an already fit organism (which didn't
bleed to death) to have still greater fitness.

How can we see whether processes such as blood clotting
actually are irreducibly complex? In support of the claim that
they are not, Miller points to experiments by Rochester
University scientist Barry Hall on the evolution of a lactose-
utilizing system in *E. coli* bacteria, which could be considered
to be irreducibly complex. But Behe notes that Hall's experiment didn't wipe out a multipart system and evolve it from
scratch. Rather, Hall merely deleted one component of the
system and provided the bacteria with alternative nutrients
that allowed them to function even with one gene removed.
Thus, Behe argues that the experiment offers little support
for Miller's claim that this particular irreducibly complex
system could evolve. (This objection seems illigitimate, however, since one needs to start with a functioning precursor of
some kind to demonstrate evolution to more complex systems.)
Furthermore, Behe wants more than a description of a series of
steps that *might* have occurred in Darwinian fashion, but also
clear evidence that each of the steps would have actually taken

51   the organism in the direction of increasing fitness. (It begs the question to insist that they must have, since otherwise the organism would not have evolved to its present state.)

Behe has actually proposed an experimental test that could settle the matter of whether irreducibly complex structures can evolve, but the proposed test is very likely not feasible. He wants experimenters to start with some specific functioning precursor structure (say a bacterium without a flagellum) and evolve a flagellum in the lab. The problem with that scenario is that the exact nature of the precursor organism can only be guessed. There is no reason to suppose that the ancestor of bacteria having flagella closely resembles present-day flagella-free bacteria, since they have found other ways to survive without one.

Moreover, even if an exact living precursor organism could be identified, neither the organism, nor its earlier environment (and competitors) may exist any longer. Furthermore, even if they did exist, the exact course taken by evolution is random, so that there is no guarantee that the same precursor would lead to what is now found, and do so in a time that is less than the lifetime of the researcher! Lacking an actual experimental creation of a complex structure in the lab, the best that one can do is to create scenarios for the path evolution might have followed, and Behe can always find some degree of hand-waving in such scenarios that leads him to question their realism.

If Behe's idea of irreducible complexity has a generality beyond the specifics of biochemistry, as he has claimed, it should be possible to apply his ideas to manmade structures for which it might be easier for us to evaluate the arguments on each side. Behe's favorite example of an irreducibly complex manmade object is the standard mousetrap, which consists of five parts: (1) a flat wooden base, (2) a metal hammer that crushes the mouse, (3) a spring that powers the hammer, (4) a "hold-down bar" that holds the hammer back, and (5) a catch that releases the hold-down bar. Behe notes that the separate components are clearly well matched, i.e., they fit together in a precise way,

52   and that the trap would stop functioning if one of them were removed or substantially altered, thereby meeting his definition of irreducible complexity.

How to Evolve a Mousetrap

In one sense it seems nonsensical to speak of evolving a man-made object such as a mousetrap, which everyone realizes was designed. But if it can be shown that a mousetrap can in fact be "evolved" from simpler structures, Behe's notion of irreducible complexity would become highly dubious. So, let's take up the challenge. To say that a standard mousetrap (figure 3.2) was evolved means that we can start from some much simpler object suited for catching or killing mice, and evolve the standard trap, step by Darwinian step. Thus, each intermediate step must make our evolving object better suited (more fit) to catch or kill mice.

University of Delaware biologist John McDonald has taken Behe up on his challenge to show how a standard mousetrap could evolve from something much simpler. McDonald shows how you can get to Behe's 5-part standard mousetrap starting from a series of simpler traps having 4, 3, 2, and finally only 1 single part. The series of progressively simpler mousetraps are shown in figure 3.3 taken from McDonald's web site, where they are depicted in a series of animated drawings (catching

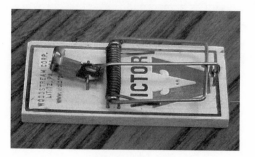

Figure 3.2 A standard five-component mouse-trap, consisting of a wooden base, spring, hammer, hold-down bar, and catch.

53 mice); see http://udel.edu/~mcdonald/mousetrap.html. (No
actual mice are harmed in the animations.)

Naturally, McDonald's simpler mousetraps (made from the
same parts as the completed version) don't work as well as
the standard 5-part mousetrap, but that's the whole point.
Complex structures can evolve from simpler precursors if the
additions make them more fit. Behe, however, has raised an important objection that throws some doubt over whether
McDonald's series of mousetraps represent a genuine evolutionary sequence. Evolution cannot take you from the 4-part
mousetrap to the final 5-part version shown in figure 3.3 in a
single step, but only in a series of steps that make small
changes in the shape of any component—at least according to
Behe's understanding of evolution. In that case, it is unclear
how those intermediate steps would each improve fitness, because although we are going eventually from one working
mousetrap to a better one, the small steps to get there lead to a
series of nonworking interim traps, which would be a non-
Darwinian process.

How can we answer Behe's objection about McDonald's sequence of traps not being part of an evolutionary sequence?
What follows is my own attempt at evolving a standard mousetrap that attempts to meet the objection. For simplicity, we'll
start with a 3-part assembly and then go the rest of the way,
rather than try to evolve the whole thing. Our starting point

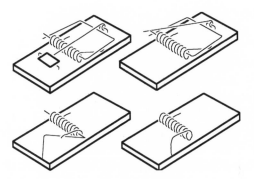

Figure 3.3 John
McDonald's 2-, 3-, 4-, and
5-part mousetraps, cour-
tesy of John McDonald.

54   structure will therefore closely resemble a standard 5-part mousetrap, but it is missing both the hold-down bar and the catch. Obviously, it is not a very good object for catching or killing mice, but it could serve this purpose to some very limited degree by being a launcher of projectiles aimed at the mice.

For example, if a number of sharp objects are placed along the edge of the hammer when it is held down by hand they could all be launched simultaneously at the pesky critters. Admittedly, such a device would be a very poor mouse killer, but it could serve that purpose to some degree in a world without traps that is overrun with mice. It is irrelevant to object that to use the device a human needs to hold the hammer back and put projectiles on it, since some sort of human intervention is needed for *any* mousetrap. Let's now imagine that small random mutations in the structure and composition of our projectile launcher occur. Most of these random mutations will either have no effect on its performance or make it worse, but a few will make it better.

For example, consider a class of mutations that slightly change the properties of the spring. Those mutations that lead to a stronger spring could make the device a more effective projectile launcher, but only up to a point, because a human hand needs to pull the hammer back. Now, think of another class of mutations, which lead to the growth of a small flap of flexible material (like skin or leather) out of the base of the trap. If this flap should arise in some places it will either interfere with the action of the spring or else have no effect on it. Those mutations will die out because they degrade fitness. However, if the flap originates just behind the location of the hand-cocked hammer it will prove very useful to a human trying to hold down the hammer in the case of a fairly strong spring.

Such a flap would improve the fitness of the projectile launcher slightly, because it would allow a human to hold the hammer down for long periods without strain. The flap also opens the door to further improvements in fitness associated with the spring becoming stronger. Once the flexible flap has

55   evolved, its effectiveness holding the hammer down by hand would become greater if it were to mutate gradually into an increasingly stiff object, and become a lightweight object of increasingly greater length, i.e., a thin, stiff bar. That improvement in fitness with bar length is a matter of simple leverage. In other words, stiffer and longer (but of lighter weight) hold-down bars permit a human to hold the hammer down with less force than do flexible and short hold-down bars.

Furthermore, this gradual evolution of the length and stiffness of the bar allows simultaneous improvements in the strength of the spring. These coupled improvements are an example of what is known as coevolution—the simultaneous evolution of two properties that feed off each other in a synergistic manner. So far, we have suggested a scenario for evolving a hold-down bar. Where does the catch, the last part of the standard mousetrap, come in?

Recall that the object we are evolving is so far an increasingly fit projectile launcher, and not yet a device for trapping mice. Let's assume that random mutations of the base again cause pieces of material to be added to it at random points, which is how we got the hold-down flap earlier. One circumstance in which such a random mutation might be useful would be if the added material caused the hold-down bar to get caught on it. That mutation would be useful (increase fitness) because then we wouldn't need to actually hold down the bar all the time waiting for the moment to release it. Instead, we could simply remove the bar from the newly evolved catch at the moment we wanted to launch projectiles toward the mice.

Still further mutations of the shape of the hold-down catch would change it from one that needed to be released by hand to an unstable catch that would be released with the slightest jiggling. At this point the device has become ideally suited to a new purpose, namely catching mice in a trap rather than killing them using projectiles—the biological equivalent of a new species. (The new "purpose" of the object should not be taken to mean that the evolutionary process that got us there was

56   purposeful. Each step in the process—each mutation—had no
goal, yet the only the fittest mutations of the object survived
over time.)

How to Ascend Mount Mousetrap

One useful way to portray the steps in the evolution of the final
5-part mousetrap described above is in terms of a concept
known as the fitness landscape. We earlier mentioned a number of properties of the evolving object, but let's focus on only
two: the length of the hold-down bar, L, and the strength of the
spring, k. Let's consider how the fitness of the device (while it is
still a projectile launcher and not yet a trap) depends on these
two quantities.

In figure 3.4, we show a possible graph of fitness (the height
of the 3-D surface) versus L and k. The peak of this surface at
point C was the point just before the catch evolved and the object became more useful to us as a mousetrap than a projectile
launcher. The reason there is a peak in the 3-D surface is that
there is an optimum k and L for the projectile launcher, everything else being fixed. In other words, if k is too small (if the
spring is too weak), then projectiles don't go very far; and if k is
too large, then we can't pull the hammer back.

Likewise if L is too small (if the bar is too short), then it
doesn't reach to the end of the hammer to hold it down; and if
L is too long, then its excess weight will slow down the hammer
when it is released. We can refer to point C as the peak of
"Mount Mousetrap." Notice that just as with climbing real
mountains, which have routes of varying difficulty leading to
the summit, there was a particular route that evolution followed to get us to the peak involving the most gradual climb.
That route started at point A and then moved to B and finally C,
all the while increasing fitness (surface height) very gradually.

The initial part of the trek up the peak A to B was a period
of increasing the strength of the spring (before the hold-down
bar evolved). At point B further increases in the strength of

57 the spring alone would not improve fitness, because the spring was getting too strong for the hammer to be held down by hand. But once a flap evolved and became a full-fledged hold-down bar, further increases in spring strength could advance in parallel, because such paired mutations lead to increasing fitness. This coevolution—simultaneous increases

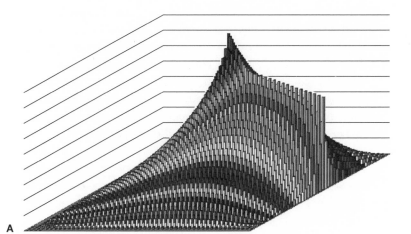

A

Fitness Contours

Spring Strength (k)

C

B

A

Bar Length (L)

B

Figure 3.4 "Mount Mousetrap" shown as a 3-D surface and a contour plot. The most gradual ascent to the summit is along path A→B→C.

58   spring strength and hold-down bar length—occurs along the path from B to C. It is the key to climbing Mount Mousetrap, because otherwise we would have to advance up some incredibly steep terrain.

The story becomes even more interesting once the object's purpose changes from being a projectile launcher to being a mousetrap. Once the object is a mousetrap (where the slightest vibration will release the hold-down bar), any changes in k and L from the peak values lead to much more drastic decreases in fitness than when the device was a projectile launcher. In fact, if L is decreased a bit and the bar no longer reaches the catch, the device becomes completely useless as a mousetrap. This means that the 3-D fitness landscape in figure 3.4 no longer applies, and instead we have a landscape such as that in figure 3.5, where a climber would be stranded on the peak with no way down.

In evolutionary terms, it is unclear how one gets to the final very fit mousetrap starting from a simpler version because the

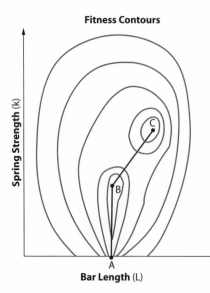

Fitness Contours

Spring Strength (k)

Bar Length (L)

Figure 3.5 A contour plot of Mount Mousetrap after the catch evolves.

59   landscape has drastically changed with the change of definition
of fitness (a better mousetrap rather than a better launcher). *In
other words, the device has at last become irreducibly complex by
changing its function in the last step of its evolution, in which the
easy route to the summit C was washed away.*

Will Michael Behe find the preceding account of how to
evolve a mousetrap persuasive? That seems unlikely. One objection might be that we started with a 3-part object rather than
an object consisting of a single part. But, one has to start somewhere. Even with a 1-part starting point (say the rectangular
wooden base that could serve as a mouse killer by being
thrown at them), we could ask what it could have evolved
from? *However, it must be stressed, for the purpose of satisfying
Behe's criteria, that our initial 3-part launcher was not irreducibly
complex, and yet the final 5-part trap was.*

Another possible objection is that during the evolution of
the trap two independent properties coevolved in tandem.
However, coevolution is a well-known biological phenomenon
that accounts for a wide range of developments both in individuals, and between individuals, such as symbiotic behavior.

Coevolution in the Development of Echolocation in Bats

In his 1986 book, *The Blind Watchmaker,* Oxford University biologist
Richard Dawkins discussed the evolution of the echolocation system that
bats use to fly in complete darkness. We cannot know what goes on in
the mind of a bat, but if we believe that most animals form visual images
of their environment, it seems reasonable that bats do the same using ul-
trasonics. Bats accomplish this feat by sending out a timed series of
high-pitched (ultrasonic) squeaks and detecting their reflections off of ob-
stacles. Submarines use a much more primitive version of the bat's
echolocation in their sonar system. Even humans have some small ability
to sense their surroundings based on echolocation, which is somewhat
better developed in the blind, because they rely on it more. The sophisti-

60   cated echolocation of bats was clearly an instance of coevolution, since
the systems for generating the squeaks and detecting and analyzing
them had to have evolved in parallel. (Coevolution does not imply that the
two components originated simultaneously, only that their improvements
over time fed off one another.) Finally, the system is an irreducibly complex one, since the various well-matched components are all necessary
for its operation, yet it was able to evolve as a result of coevolution.

Another potential problem with my mousetrap evolution
scenario is the question of what constitutes a "small" incremental change? Obviously, my scenario is unrealistic if one insists
that only the tiniest steps in structure change can occur in individual mutations. But, fortunately, real evolution sometimes
doesn't work that way. Sometimes random mutations lead to
very large changes in an organism—a calf growing an extra leg
or head, for example. More importantly, as this example illustrates, sometimes the random mutations make use of whole
components or subsystems that can be put together as building
blocks.

In fact, the evolution of useful components is probably the
real secret behind the development of increasingly complex artificial or natural systems and organisms. In the case of biological evolution, useful components can get created during the
process of gene duplication. This process, which has been observed in the lab, involves the copying of fragments of DNA sequences from one place on the genome to another. Thus, once
evolution successfully created the first neuron in some ancient
organism, the process of gene duplication would make it that
much easier to create an organism with many neurons, and
then an actual brain.

The importance of general-purpose components is very clear
in building complex man-made objects. Think of electronic circuits for a moment. Suppose we imagined trying to evolve a
computer by mimicking the principles of biological evolution.

61   We might start by generating many assemblies of one million transistors connected at random. Almost certainly the resultant assemblies would be piles of junk in virtually every case. Even after we selected the cases where the "fittest" devices did something vaguely interesting and allowed them to mutate and continued the selection process, we still might not expect anything resembling a functioning computer (or neural network) to emerge for a very long time.

However, if, in the course of evolution, starting with our random collection of a million transistors, certain simple components, such as the logic circuits for the functions AND, OR, NOT, XOR were to evolve first, then the chances of those components randomly making something interesting increases. The chances of getting something interesting and useful increase still further if those logic elements should randomly aggregate to form general purpose components at a higher level—circuits such as amplifiers, timers, and filters. Putting it differently, the chances of getting something useful are much greater when instead of random hookups of a million transistors, we had random hookups of a thousand logic circuits or, better still, a hundred microprocessors.

Artificial Life and the Evolution of Robots

Artificial life is sometimes described as life as it might be if it were not constrained by the requirements of biochemistry. Simulations are often used in this field to evolve structures and imaginary organisms in the computer, based on the principles of Darwinian evolution. Here's a very brief description of how to use an evolutionary algorithm on the computer.

Start with a random population of simulated organisms having a range of properties, such as number of parts and the shapes and sizes of each part. Assume that these properties are specified for each organism by a particular sequence of genes. Choose pairs of organisms at random and allow them to

62 "mate" and produce offspring organisms. This mating process mimics sexual reproduction in nature in that the gene sequences of the two mating organisms are recombined or scrambled to find the gene sequence for the offspring. (See figure 2.4 for a schematic illustration of the recombination process.) In addition to being recombined, some bits of the genetic code are altered at random to mimic the mutation process. Most importantly, the selection of which organisms are allowed to mate is based on some measure of their "fitness." The whole process of selecting the fittest artificial organisms and allowing them to mate to produce offspring is then repeated for many generations. The result is a population that becomes increasingly fit over time.

The type of artificial life simulation just described differs from real biological evolution in at least one key respect. In biological evolution it is assumed that an organism's fitness for its particular environment together with chance will determine its reproductive success. In the field of artificial life, however, fitness is instead determined by assigning some arbitrary goal, such as being able to move with the greatest speed. The computer program then generates simulated organisms and allows them to mutate and reproduce. In effect, the program rewards those organisms that have greater fitness by allowing them to have more offspring. There are many reasons for wondering whether these artificial life simulations can tell us anything useful about the question of whether irreducibly complex structures can be evolved in nature. Let us consider some of the questions and see how they can be answered.

Problem 1. How to judge fitness. Rewarding artificial organisms for some desired property, such as how fast they can move, is different from what happens with real organisms, where all that counts is survival, or at least survival to reproductive age. The simulation can be made more true to life by having the desired property (say speed of movement) directly tied to survival, for example, by having the artificial organism feed as it moves through its environment. For greater realism,

63   we could include predator organisms and various environmental challenges to survival.

Problem 2. How to go from the computer to the real world. In some artificial life projects, such as the automatic design and manufacture of robotic life forms at Brandeis University, led by Jordan Pollack, the computer simulations are directly tied to real world objects, in this case actual robots (figure 3.6). One starts with some set of designs chosen at random that are used to construct real robots automatically. The fitness of each of the real robots is evaluated for desired property (say speed of movement). The robots are then paired, and their designs "mated" (taking parts from one and parts from the other), with more matings taking place, depending on the fitness. This process mimics what happens in biological evolution.

Problem 3. How to avoid design being built in. If we were to try to evolve extremely sophisticated systems starting from very sophisticated components it could always be argued that the design was present at the outset. Usually, artificial life sim-

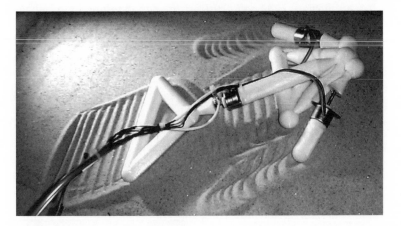

Figure 3.6 Photo of a robot designed through an evolutionary algorithm, supplied by Jordan Pollack. © 2000 Lipson and Pollack, printed with permission.

64 ulations avoid this problem by starting with fairly simple components. For example, in the Brandeis lab, robots are automatically designed using only fixed-length bars that can connect to one another, actuators that can vary the angle between connecting bars, and artificial neurons as building blocks of a control circuit "brain." To quote from the web site for the project (www.demo.cs.brandeis.edu/index.html):

> Starting with a population of 200 machines that were comprised initially of zero bars and zero neurons, we conducted evolution in simulation. The fitness of a machine was determined by its locomotion ability.... The process iteratively selected fitter machines, created offspring by adding, modifying and removing building blocks [bars and neurons], and replaced them into the population.... This process typically continued for 300 to 600 generations. Both body (morphology) and brain (control) were thus coevolved simultaneously.... Typically, several tens of generations passed before the first movement occurred. For example, at a minimum, a neural network generating varying output must assemble and connect to an actuator for any motion at all.

Problem 4. How can artificial life say anything useful about real biological evolution? The beauty of artificial-life simulations is that they clearly illustrate the creative side of the process of natural selection. The natural selection process is not just one of weeding out unfit organisms, but it allows more successful life forms to emerge by rewarding the rare beneficial mutations. The "reward" here is the same as in real biological evolution, namely the fitter organisms get to pass down their traits to the next generation of artificial organisms. One interesting feature of the Brandeis experiments is that as a robot evolves, its fitness (here locomotion speed) seems to improve in discontinuous jumps rather than perfectly smoothly from one generation to the next. These rare sudden jumps in fitness attest to the rarity of significantly beneficial mutations. They also may signal the development of a new "species." These sudden

65 jumps in fitness may mimic what happens with the formation of species in real biological evolution, according to both the fossil record and Stephen Jay Gould and Niles Eldredge's theory of "punctuated equilibrium."[11]

"Punk-eek"

The idea of punctuated equilibrium was suggested by Gould and Eldredge based on the observation that the fossil record seems to show long periods of stasis with intervening bursts of change that occur during "instants" of geological time—which could be thousands of years—when new species are created. At first blush this idea of Gould and Eldredge seems to run counter to the basic idea of Darwinian evolution, which demands that all change occur through a series of slight successive variations, and never in sudden leaps. In fact, some creationists have misconstrued the idea of punctuated equilibrium to claim that new species appear quite suddenly, as would be expected if they were created, not evolved.

Gould, however, has emphasized that the punctuated equilibrium theory is definitely not in conflict with the ideas of Darwinian evolution by natural selection.[12] The apparent conflict with Darwinian evolution can be resolved by noting that the many slight successive variations required for new species to form could take place in times that are short on a geological timescale, especially during periods of drastic environmental change. It is precisely during these periods when many species become extinct and new ecological niches are opened that we would expect to see rapid changes in the survivors, and new species being formed.

Artificial-life simulations include a number of similarities with what is observed in real-world biological evolution, including the spontaneous development of new species, coevolution, and useful components into higher-level structures. With regard to speciation, for example, the Brandeis project has

66   evolved a variety of differently locomoting robots that can travel across a flat surface. If you wish to see some very interesting "species" of such creatures, including "Sophia—the walking star" and the "giant serpent," see the animations at www.demo.cs.brandeis.edu/pr/evo_design/genobots2d.html.

Species form in these artificial-life simulations for same reason that they form in nature. Each many-generation run of the simulation is essentially reproductively isolated from any other run. Thus, even though each run might start with the same initial random population, differences between runs steadily accumulate due to natural selection and random mutations.

Problem 5. How to evolve a specific organism. One problem with all artificial-life simulations is that, as in real biological evolution, it is not possible to start with an end result that you want to achieve and have any assurance you can get there from some given starting point. Evolution, which depends on random mutations, may lead to improvement in fitness of a certain kind, but the exact resulting organism or structure that results will be unpredictable and cannot be specified ahead of time. In other words, we could not start with a specific species, such as "Sophia—the walking star," and expect to be able to evolve it in any given run of the simulation. In much the same vein, we should not expect to be able to start with green slime and see homo sapiens after a few billion years if we could run real evolution over again. Quite apart from the preceding problem, however, the creatures that evolve in the artificial-life simulations probably do justify the label of irreducible complexity, according to Michael Behe's definition of the term, namely they have many closely matched parts that are all required for the creatures to function.

In summary, Behe's criteria for distinguishing whether a structure was designed and could not have arisen from blind evolution does not seem to be warranted, because it assumes that evolution can proceed only through successive tiny modifications in one property. In fact, however, there are a number

67   of evolutionary mechanisms that specifically exist so as to generate large changes now and then.[13] It appears that evolution through natural selection can take place by assembling earlier-evolved building blocks, by sometimes changing the function or "purpose" of the structure midstream, and by allowing for coevolution of mutually interacting properties. Essentially, Behe is being inconsistent when he insists that irreducible complexity shows itself in the biomolecular structures unknown in Darwin's day, but then relies only on the old-fashioned limited view of the evolutionary process. If Behe's irreducible complexity criterion cannot reliably detect the presence of design, might there be some other method for recognizing when a structure or process could not have evolved naturally?

William Dembski's Filter

William Dembski, an associate research professor in the conceptual foundations of science at Baylor University and editor of the 1998 book *Mere Creation,* has proposed a method of recognizing design based on what he characterizes as a simple algorithm or "explanatory filter." Dembski claims that his filter allows any event to be placed into one of three categories— "law," "chance," or "design"—based on its likelihood of occurrence. According to Dembski, events having a high probability of occurrence, given known antecedent circumstances, are in the "law" category. For example, when released from your hand a ball will (almost) always fall to the floor, due to the law of gravity. The "almost" qualifier is intended to take into account freak circumstances, such as a sudden upward gust of wind, which could alter the outcome.

Events not in the law category might have an intermediate probability of occurrence—sometimes they occur and sometimes they don't, given prior circumstances. Such events Dembski attributes to chance. For example, let's say you opened a box of 1000 coins and found 507 of them heads up, which is not a terribly unlikely outcome. (The most likely out-

68   come of exactly 500 heads occurs 2.5 percent of the time on the
basis of chance.) If someone predicted 507 heads *ahead of time*,
you might be impressed at the accuracy of her guess, but not so
impressed that you would conclude that either she was psychic
or had arranged things ahead of time. But, with no advance
prediction, there is nothing particularly surprising about finding 507 heads out of 1000 coins, and you would probably conclude you had observed a chance event.

Dembski's third category of events are those that have been
specified ahead of time and have a small probability of occurrence. In such cases, both law and chance can be ruled out as
possible explanations, and the event can be attributed to design. Thus, if we specified in advance that we would find all
1000 coins heads up, and found precisely that outcome on
opening the box, the result could not be explained without human interference of some form. (But, of course, that inference
requires that we know the coins have heads only on one side.)

Can Dembski's filter reliably sort out designed events in the
real world, particularly events signaling design in the formation of biological structures? Clearly, some designed events
("false negatives") would escape detection by his filter. (Someone who actually arranged the 1000 coins to have 507 heads up
would have his efforts unrecognized by the filter, since 507
heads is not that unlikely to occur by chance.) But we need not
be concerned here about the false negatives, only the possibility
of false positives, i.e., those cases where we infer design when
there really isn't any. Does Dembski's filter allow false positives
to slip through?

In the coin example, the answer would clearly be no.
Dembski argues that in a wide range of other real world examples the answer is also no, including SETI (search for extraterrestrial intelligence), cryptography, forensics, and archeology. If
an archeologist found an ancient artifact that bore a strong resemblance to a tool or other man-made object, he could reasonably infer it was designed even if its purpose might be obscure.
Likewise, from the physical evidence, a forensic detective
might infer that a body was the result of a murder (an act of

69 design) rather than an accident, based on such evidence. Of course, in many cases the evidence would be convincing only if it were one piece among many. The body of a left-handed victim found holding a gun in his right hand, for example, would arouse suspicion, but death could possibly have been the result of a shooting accident or suicide.

Ultimately, while Dembski's filter is applicable to many areas, the real question is what it says about whether biological organisms may have been designed. One major problem in applying the filter is that evolutionists and intelligent designers will have extremely different estimates of the probability of particular steps being taken in an evolutionary sequence leading to a certain complex structure. For example, if one restricts Darwinian evolution to a sequence of baby steps in changes in function and shape of an organism, major changes leading to new functionality become exceedingly unlikely. But, if we include coevolution, development of functional higher level components, and sudden changes in function of structures, then the probabilities are likely to be very much greater.

Changes in Function during Evolution

There are many examples of structures changing their function during the course of biological evolution. One example is the three small bones in the mammalian middle ear, two of which appear to have evolved from bones that were originally located in the lower jaw of our ancestors.[14] The appearance of wings in insects about 330 million years ago offers a second example. Genetic evidence shows that wings evolved from articulated gill plates on the limbs of aquatic ancestors.[15] An example of a change in function at the genetic level involves the genes responsible for creating the transparent cells in the lens of the eye. According to the fossil record, the first eyes evolved in a primitive fish around 530 million years ago. Apparently, this occurred when genes used to manufacture certain metabolic enzymes mutated and developed a second function, allowing them to "moonlight" and produce proteins active inside developing

70   eyes.[16] For a fourth example of the process of how evolution works by changing function, consider the mechanism of blood clotting, deemed by Behe to be irreducibly complex. The proteins used to clot blood appear to have evolved from proteins used in the digestive system. Russell Doolittle has shown how the genes for these digestive proteins were copied and modified during the course of evolution to serve a new function.[17]

Dembski's filter also runs into problems when we try to judge the probability of a nonbiological system arising by chance. Consider the example of Mount Rushmore. Suppose you had never heard of Mount Rushmore, and stumbled upon this mountain with the faces of four presidents carved into it. You wouldn't have the slightest doubt that the structure had been designed, based on the extreme improbability of getting those shapes by chance. But now suppose you stumbled on a version of Mount Rushmore carved by aliens for their leaders, and you had no clue as to what the aliens actually looked like before seeing the mountain or what body part of their leaders they preferred to display on their monuments. Let's assume further that you had no clue as to what the natural topography of the alien planet looked like. Do you really believe that under these three conditions you could distinguish an alien Mount Rushmore from a natural terrain feature with high confidence? The same could be asked about intelligent aliens being shown pictures of our Mount Rushmore before they had seen humans or Earth's natural terrain.

If you still doubt that many observers, including some scientists, might be fooled into thinking they have seen evidence for design in natural objects, consider these two examples from history. Astronomer Johannes Kepler once claimed that the craters on the moon were designed objects, because he apparently was unaware of the mechanism of cratering by meteorite impact—though it is unclear why he didn't consider that they were produced by volcanoes. Similarly, the astronomer Percival Lowell believed in artificial canals on Mars

71   (presumably built by Martians), based on his mistaken observation of a network of connecting lines on the planet's surface.

Lowell was deceived by his eyes and his preconceptions, while Kepler was deceived by his lack of imagination, but both applied inductive logic to draw their false conclusions. In effect, they observed that any time they had ever seen very large perfect circles or straight lines drawn on the ground, those shapes were put there by design. Therefore, they improperly inferred a designer in these two extraterrestrial cases as well. Still more recently the "face" on Mars has been cited as evidence of Martians. This last example is the most egregious of the three (see chapter 2), but the other two show that even serious scientists can be fooled into seeing design when it is not present. How much more likely is one to fall into that trap when the design inference falls comfortably into one's religious view? Ultimately, I believe Dembski's filter is of no real value in deciding whether biological systems may have been designed.

The Anthropic Principle and Intelligent Design

The anthropic principle was first proposed by Brandon Carter in 1974.[18] Although the principle comes in various forms, the "weak" version, as enunciated by physicists John Barrow and Frank Tipler, says that

> The observed values of all physical and cosmological quantities are not equally probable, but they take on values restricted by the requirement that there exist sites where carbon-based life can evolve and by the requirement that the Universe be old enough for it to have already done so.[19]

The kinds of quantities the principle refers to include various constants, such as G, which determines the gravitational force between any two masses. Thus, the anthropic principle notes that if the strength of gravity were either slightly greater or less than its actual value, life (at least carbon-based life) could not have evolved. With a slightly larger G, it turns out that only red dwarf stars would exist, which are too cold to permit

72   planets in a life-sustaining habitable zone around them. Similarly, if *G* were slightly smaller, all stars would be blue giants, and they would live too short a time for the evolution of life. *G* is just one of many constants, all of which need to be "finely tuned" to permit life to exist now in the universe.

Some observers see in this fine tuning of the constants of nature clear evidence that our universe was designed. In other words, just based on random chance, it would seem highly improbable that the universe we live in had just the right values of constants to allow life to exist. On the other hand, it is also true that given that we are alive, it is impossible that the constants could have taken on values that would make life impossible. In that sense, the probability of finding the constants of nature taking on their finely tuned values is 100 percent. Conceivably, there are other sterile portions of the universe where the "constants" of nature do take on other values.[20] But in those sterile regions there wouldn't be any observers to wonder why the constants of nature took on the values they did!

A similar paradox concerns the seeming improbability of our own *individual* existences. Suppose that in ancient times only one in two babies born survived to reproductive age. We might then calculate the probability of 1000 of your direct ancestors surviving to their reproductive years to be $1/2^{1000}$, which is such a tiny number that there is almost no chance you would have been born. Obviously, since we know you were born, such a calculation is fallacious for exactly the same reason that it is incorrect to say the fine tuning of the constants of nature is extremely improbable. (If they were not so fine tuned we wouldn't be here thinking about it.) In any case, the anthropic principle serves again to remind us that a given observation can be interpreted as either a highly improbable event (and possible evidence for design) or one that could not have been otherwise, which therefore needs no explanation.

William Dembski has also proposed a second strategy based on information theory for recognizing intelligent design. He notes that we can often recognize the presence of an intelligent agent at work and distinguish it from either chance or natural

73   law by seeing the presence of "complex specified information." For example, if SETI researchers were to detect a signal from space containing the first 1000 digits of pi, they would have strong reason to believe that it was a real signal because of its complexity and specificity, especially if that signal were repeated. On the other hand, the detection of a more monotonous string of beeps equally spaced in time is more likely to be of natural origin. (However, it should be noted that Jocelyn Bell, the researcher who first detected the string of beeps from the first pulsar, briefly did consider that it was an artificially created extraterrestrial attention-getter.)

Dembski goes on to observe that the amount of information is never increased or created when either chance or well-defined lawful processes act. In other words, in the presence of well-defined completely predictable processes any information in a system was either contained in it from the beginning or supplied from the outside. Similarly, in the case of random chance, information can be degraded, such as with the addition of noise to a signal, but information is never increased by the action of chance alone. Dembski therefore argues that if the information in living systems becomes more complex over time, it is a sign that something other than chance or natural law—specifically design—must be operating, based on his idea of "conservation of information."

What Dembski fails to recognize is that although the action of either chance or natural law (completely predictable processes) will not increase the complexity of information, their *joint* action can have precisely that outcome. Examples of this type of creative interaction are numerous. They include the use of natural selection to evolve complex forms of artificial life, computer programs, concepts and behaviors in humans, and forms of real organisms.

When a robot learns to move toward the light—to take one simple example—that algorithm was not fed to it by its original programming; rather like a human, it learned by trial and error while interacting with its environment. The robot's fitness can improve over time if the robot starts with a random algorithm

74 that mutates, and more successful mutations are rewarded through natural selection. In contrast to claims of some creationists, "success" is here not defined by a comparison between what the robot "sees" and some preprogrammed target defined by the human programmer. Rather, those robots whose initially random controller circuits lead them to the light sooner are rewarded.

Ultimately, Dembski's method of recognizing design based on the complexity of information, while interesting, seems no more useful as an "ID detector" than his three-part filter.

When Is a Theory Scientific?

According to philosopher of science Karl Popper, the primary characteristic of a theory that claims to be scientific is that it be *falsifiable,* i.e., testable through observation, and capable of being proven wrong. Ideally, the means of the testing should be clearly specified and be capable of being carried out by anyone with the appropriate technology. In addition, either the theory should make some specific linkages clear between phenomena previously thought to be unrelated or it should make some predictions of quantities that can be measured without any subjectivity involved.

What are some ways theories might fail to meet these criteria? Falsifiability is the most important of the criteria, so we would not be particularly impressed by a theory that could not be tested by experiment or observation. (We include both experiment and observation, because in some fields, such as astrophysics, all we can do is observe the cosmos, since no actual controlled experiments are possible, e.g., assembling a star or a galaxy in the lab.) It is nonsense to claim, as some creationists do, that we can only have confidence in theories testable in lab experiments. We can have very high confidence that the stars are made of the same elements found on Earth, for example, based on observations of their spectra.

75   On the other hand, falsifiability is sometimes less than clear-cut, because some theories can be tested easily using present day technology and knowledge, while others, such as string theory, may require great advances in technology before they can be tested. Assuming that a theory can be tested, the specificity of the test is as important as its feasibility. For example, some theories claim that the speed of light is changing over time. However, unless the theory gives the *rate* at which the speed of light is changing, it cannot be disproven, because measurement uncertainties will always leave room for some level of variation over time. In other words, if no variation in speed is found, all that an experiment could do is place an upper limit on the extent of its variation over time. In addition, it is vital that the test of a theory leave little room for subjective interpretation. A theory that plants have feelings that can be observed only by particularly empathetic researchers and that these feelings are negated when plants are studied by skeptical investigators should not impress us very much.

Ideally, scientific theories need to predict *new* phenomena. They might also provide numerical values of known quantities and relate previously unrelated quantities. For example, a theory that calculated the masses of the electron and proton exactly in terms of other known fundamental constants would be fascinating, even if it didn't make any predictions. But theories that merely "explained" very well-known facts about the world would not be particularly impressive, e.g., a theory that the speed of light should not change over time or that pigs should not be able to fly.

How does the theory of intelligent design measure up based on the preceding criteria? Concerning predictive power, zero new phenomena or numerical results of any kind have been made by the theory. The theory also does not tie together any known phenomena that were thought to be independent. In contrast, the theory of evolution explains much, including the manner in which new species evolve, the common ancestry of all life, and many otherwise puzzling behavioral adaptations of

76   organisms. Regarding intelligent design, I believe it is testable, although perhaps not in the way that its adherents think. Michael Behe has suggested that one way to test ID and prove it wrong would be for researchers to evolve a flagellum in the lab starting with an organism that didn't have one. For reasons given earlier, I believe such an experiment could not be carried out, so Behe is not proposing a feasible test.

Even if by happenstance a bacteria flagellum could be evolved in the lab in the lifetime of the researchers, all that Behe would concede is that that particular structure (and others of less complexity) could be evolved. But the question of intelligent design of other more complex structures would still remain open. Behe's proposed test also puts the burden of proof on an opposing theory (evolution) to prove itself in the lab, before he would accept that ID might be in trouble. In fact, evolution *has* been demonstrated in everyday life, such as when successive generations of bacteria evolve resistance to antibiotics, thereby becoming increasingly fit. These demonstrations are dismissed as merely examples of microevolution (evolution within a species) by ID adherents, but the distinction between macro- and microevolution is not so clear for bacteria.

So, what would be an appropriate test of ID? Putting the burden of proof on ID and doing tests that we know are feasible means that a proper test is the converse of what has Behe has suggested. People wishing to test ID should examine a wide range of biochemical structures and functions evolved in the lab, and show that such structures are never irreducibly complex. One example to the contrary would disprove ID. I suspect that little along these lines has been carried out, and that ID adherents might not be particularly keen to make these tests. If one wants to include artificial life, however, it could be argued that ID has already been falsified, because many of the end products do involve irreducibly complex structures. Arguments about whether the design was built in at the start and whether the designer really kept his hands off or not during the evolution suggest that such judgments will probably forever make the test of ID a subjective matter.

77   In summary, the theory of intelligent design is not a scientifically valid alternative to evolution, and, further, it is probably not in the domain of science. The various means proposed to recognize when something has been designed (irreducible complexity, Dembski's filter, information complexity, and the anthropic principle) each have their shortcomings, and they cannot be counted on to distinguish designed from natural systems with confidence. Moreover, ID makes no predictions, explains nothing about living systems that was not known previously, and may be untestable—at least in a manner that is not open to subjective interpretation and is feasible within a specified period of time.

I'd give the idea that intelligent design is a viable scientific alternative to evolution a 3-flake rating. We must, however, make one concession to intelligent design. In the future, given our understanding in genetics, it is likely that changes in the structure of organisms due to human intentions will dwarf those brought about by natural selection. The future evolution of organisms is likely to be driven by *our* design.

4 Are People Getting Smarter or Dumber?

IN HIS DARK MOMENTS (usually after grading a physics exam) this long-time college professor sometimes wonders if there has been a general deterioration in human reasoning ability. Fortunately, on most days this feeling is fleeting. On other occasions when I am pleasantly surprised by a particularly insightful student comment I ask myself the reverse question: might people actually be getting smarter? I'm unaware of any polls on either the smarter or dumber question, but I imagine that when young and old people look at one another across the generational divide each might more often tend to regard its group with greater favor in terms of intelligence. Just for curiosity I did a web search for the words "people getting dumber" and "people getting smarter" on my favorite search engine and came up with the following results:

Phrase	Exact Phrase	Words Nearby
"People getting dumber"	29 hits	25,700 hits
"People getting smarter"	40 hits	180,000 hits

If we tried to conclude anything based on this meaningless exercise, it probably would say more about whether the author is getting dumber than anything else, but I did find the 7-to-1 ratio in favor of "smarter" somewhat surprising. Maybe the word "dumber" has too many negative connotations and tends to be avoided nowadays. Anyone looking for evidence that people are getting dumber can find many examples of stupidity that seem to be on the rise in today's world. One of my candidates would be the increasing number of people who choose to play the lottery, but only when the jackpot reaches $20 million, believing that anything less won't make a major change in their lives.[1]

79   One compilation of stupidity by individuals that makes
for quite interesting reading are the "Darwin Awards." A
compilation of these awards can be found at the web site
www.darwinawards.com and in the book *The Darwin Awards:
Evolution in Action.*[2] The Darwin Awards, which are simultaneously sad, cruel, and funny, commemorate those who have improved the human gene pool by removing themselves from it.

Contrary to popular belief, however, Darwinian evolution
says nothing about whether intelligence should increase or decrease over time. There is little doubt about the actual increase
that has occurred when we compare the relative intelligence of
humans with that of the extinct species from which we have
descended. But that development shows only that in the environments in which these species found themselves greater intelligence had survival value. It is quite possible to imagine a
different planetary history and different environments in
which intelligence would not have had great value. (One can
easily imagine postapocalyptic futures in which keen senses,
brute strength, ruthlessness, and the ability to withstand hard-
ship or extreme heat would have much greater survival value
than intelligence.)

Any serious attempt to try to learn whether people are getting smarter or dumber over time immediately runs into at
least four difficult questions:

- What do we mean by intelligence?
- How can intelligence be measured?
- Which people are we talking about?
- What time interval are we considering?

Sometimes half-jokingly it is said that intelligence is what intelligence tests test. While that circular definition is not particularly helpful in clarifying the meaning of intelligence, it is not
entirely useless either. If the ability to score highly on intelligence tests correlates highly with real-world abilities that are
normally thought of as representing examples of mental ability,
the tests do acquire a degree of credibility. After all, a similarly
circular definition of time as "that which a clock measures" was
instrumental in leading Albert Einstein to his theory of relativity.

80   Nowadays, it has become fashionable to note that there are many kinds of intelligence and that abstract reasoning ability no longer should be regarded as the sole or even the primary measure. According to Howard Gardner's theory of multiple intelligences there are seven types of intelligence: verbal/linguistic, musical, logical/mathematical, spatial, body/kinesthetic, intrapersonal (e.g., insight, metacognition), and interpersonal (e.g., social skills).[3] Only two of these seven forms (verbal/linguistic and logical/mathematical) are tested on conventional IQ tests. Given Gardner's theory of multiple intelligences, one might regard a possible loss of reasoning ability and a corresponding gain of "emotional intelligence" as simply a shift from one kind of intelligence to another, rather than a dumbing down of society. Notwithstanding Gardner's multiple intelligence theory, however, in this chapter when we address the question of whether people are getting smarter or dumber over time, the primary focus will be on verbal/linguistic and logical/mathematical intelligence—which are measured by IQ tests.

The two issues of "which people?" and "which time interval?" listed among our four earlier questions are also important in deciding whether people are in fact getting smarter or dumber, because we may get very different answers to our questions, depending on the choices we make. For example, when we ask if people are getting smarter or dumber, are we considering a particular subgroup—college students, Americans, people living in the developed nations—or all humanity? Similarly, if we consider an interval of time covering the last two million years, there can be little question that people have gotten smarter, based on a tripling in brain size and the development of language. The issue is much more problematic though when we consider time intervals of decades or centuries.

Are People Getting Dumber?

One possible indicator that Americans are getting dumber could be a continued decline of the education system. Nowadays, we

81 hear many complaints about public education in the United States and its sorry condition compared to the past. For example, consider the following eight recent quotations from a variety of leaders in government, business, and education:

- "The quality of public education seems to have declined, and schools are not up to the task of readying young people for the challenges of the next century. An apparently watered-down curriculum ensures that all students, regardless of whether they have mastered necessary skills, can graduate."[4]
- "Public education has put this country at a terrible competitive disadvantage.... If current ... trends continue, American business will have to hire a million new workers a year who can't read, write or count."[5]
- "There is indisputable evidence that millions of presumably educated Americans can neither read nor write at satisfactory levels."[6]
- "A third of ninth graders ... read at only a second or third grade level because phonics had been abandoned."[7]
- "We have simply misled our students and misled the nation by handing out high-school diplomas to those who we well know had none of the intellectual qualifications that a high-school diploma is supposed to represent."[8]
- "Education faces a serious crisis.... We will suffer the consequences of our present neglect of education a generation hence."[9]
- "The results [of a recent survey] revealed a striking ignorance of even the most elementary aspects of United States history."[10]
- "40 percent of high school graduates could not perform simple arithmetic or accurately express themselves in English."[11]

Actually, I lied. These are not all "recent" quotations. In fact, their dates are, respectively, 2001, 1995, 1974, 1961, 1958, 1947, 1943, and 1992. These quotations (along with other similar ones dating back to 1896) appear in the book *The Way We Were* by

82   Richard Rothstein, the author of the first quotation.[12] In fact, Rothstein doesn't really believe the claim made in that first quotation, and instead is making the point that many observers throughout our history have always decried the sorry state of current education compared to some better imagined past that didn't really exist. Clearly, such expert opinion cannot tell us very much about whether educational standards really have declined, let alone whether this hypothesized decline is a reflection of people getting dumber over time.

What Changes Have Occurred in U. S. Education?

If we cannot trust "expert" opinion on the changing state of education in the United States, we can examine some measurable trends. For example, one study has shown that the vocabulary in textbooks has been dumbed-down over a 30-year period by two or more grade levels.[13] One other recent study (and this one *is* recent!) on the state of middle school science texts found an incredible number of serious factual errors in all the widely used texts. Apparently, few if any science educators or scientists were involved in the actual writing of these books, and their publishers have little incentive either to get the science right the first time or to make corrections later.[14] There seems to be no question that the process of creating science texts today involves much greater attention to "glitz," multiculturalism, and student involvement, and much less attention to the actual science than used to be the case.[15]

In fact, many educators today are quite dismissive even of the need to have students learn specific information, instead focusing much more attention on the need to think critically. For example, according to Terry O'Banion, former President of the League for Innovation in the Community College, "Learners no longer have to store information in their heads; they are free to analyze, integrate, solve, apply—in other words, to learn by doing something with information."[16] Although one can only applaud O'Banion's emphasis on the importance of critical thinking skills, a critical thinker entirely lacking information in

83 his or her head seems to be an oxymoron—or maybe just an ordinary moron.

The changes in educational practice over the past 50 years include a much greater attention to collaborative active learning and developing critical thinking skills, at the expense of learning specific facts. These developments seem laudable, as long as the skills being taught are not just critical thinking in name only and specific facts are not neglected entirely. For example, it probably is not that serious if people are unaware that water boils at 212°F. But it is somewhat depressing that, according to one poll, over a third of Americans believe that dinosaurs and humans coexisted on Earth[17] or that many recent Harvard graduates seem to be unaware of why the climate is warmer in summer than in winter.[18] Likewise, it is dismaying that many recent MIT graduates were unable, when challenged, to connect a single wire to a battery and a flashlight bulb so as to light the bulb.[19] (These two observations regarding Harvard and MIT graduates were based on a sample of videotaped informal interviews with students on their graduation day that have received wide attention in the media.) The dinosaur example may reveal lack of knowledge more than critical thinking skills, but the light bulb example probably reveals both deficiencies.

Sometimes it is argued that our increasingly complex technological society is de facto evidence that people cannot be getting dumber—*somebody* must be inventing all the high-tech gadgets that surround us. Indeed, on a per person basis, Americans are patenting more inventions today than at any time since the mid-1970s.[20] But, there are many problems with such arguments and statistics based on technology. First, many people would be loathe to argue that a culture lacking modern technology is in any real sense "dumber" than one that has it. Second, the statistic about more U.S. patents now than at any time since the 1970s ignores a very pronounced drop in numbers of patents that preceded the mid-1970s, and also the many other factors (such as changing laws) that affect the numbers of patents.

Third, many of today's educational elites apparently have little deep conceptual understanding of some of the basic science behind today's technology, as exemplified by the aforemen-

84   tioned surveys of Harvard and MIT graduates. And, finally, it might be argued that scientific knowledge seems to be found in a decreasing number of heads in the United States, based on declining numbers of students majoring in the *non*biological sciences in college. For example, in 1999 the number of students graduating with bachelor degrees in physics was at a 40-year low, and degrees in mathematics, computer science, and engineering have also been down during the last 15 years.[21]

On the other hand, none of these data should lead to the contrary conclusion that Americans are actually getting dumber. Students choose majors for all sorts of reasons. For example, the decline in the number of majors in the various sciences and engineering fields needs to be evaluated in the context of students choosing competing majors, such as biology, which have risen in recent years. Moreover, we shouldn't draw too sweeping a conclusion from the inability of some Harvard or MIT graduates to answer a few specific basic science questions. Those failures might indicate that the de-emphasis of education on specific facts has been carried too far, but it could also be the result of somewhat tricky questions or certain understandable student misconceptions that may have arisen from poor teaching.[22]

In addition to a de-emphasis on specific facts and written text, education today makes much greater use of technology and multimedia, including TV and computers, than in the past. Whether that particular trend represents a "dumbing down" depends on your point of view. One might, for example, argue that increased levels of skill involved with the processing of visual information more than offset a possible loss of skill relating to written text. But, it is very likely that a decline in vocabulary has occurred in recent years as the number of hours children watch television (and surf the Internet) has grown.[23]

What about Trends in Standardized Test Scores?

What has been the measurable impact of these educational trends on test scores, including the widely used SATs, taken by most college-bound high school seniors? For a time, beginning

85 around 1963, scores on the SATs went into a precipitous decline—particularly the verbal part of the exam, which dropped 50 points over the next 15 years, never to recover. The cause of that decline has been unfairly attributed to dumbing-down of high school curricula. Almost certainly, however, the main cause of the post-1963 SAT decline has been the expansion in the pool of those high school students going on to college.[24]

This expanded college-bound pool began to include increasing percentages of poorly prepared students. We can check that falling SAT scores were largely the result of an expansion in the college-bound pool by looking at trends on the PSAT exam. (Whereas the SAT is taken only by college-bound students, the PSAT has been taken by nearly all high school juniors since 1964.) The trend in PSAT scores shows no significant variation in time, and certainly no drastic 1970s drop.[25]

One other widely used indicator of educational achievement in the United States has been the National Assessment of Educational Progress (NAEP) exams given to students in grades 4, 8, and 12 to test their proficiency in reading, science, and mathematics. The average scores on the NAEP reading exams show little change over the period of the last 30 years, while those in math and science have shown significant increases.[26] Considering all the test data together, there seems to be little support for the view that American education has been dumbed down over the last 20–40 years, although the post-1963 democratization of the college-bound pool may have negatively affected the content and standards of post-secondary education. (For example, over the last several decades there has been a significant growth in remedial courses. Consider that in community colleges 58 percent of math courses were remedial in 1995, but in 1970 the number was only about a third.[27])

Has There Been a Rise in "Weird" Beliefs?

Being in touch with reality can be considered the hallmark of sanity, clear thinking, and potentially intelligence—although we are all acquainted with individuals who have some really

86   weird ideas although they are otherwise highly intelligent. Arguably, some weird or strange beliefs are indicative of less than sound judgment and perhaps a less than critical ability to examine evidence. Has the prevalence of weird beliefs risen over time? Table 4.1 shows some results from Gallup Polls for three different years decades apart.[28]

Many of the beliefs listed appear to be held by a surprisingly large and, in some cases, an increasing number of Americans, based on the three polling years considered. Moreover, in one poll, an astounding two-thirds of those surveyed claim to have had ESP (extrasensory perception) experiences themselves.[29] Calling the beliefs "weird" reflects the judgment of most mainstream scientists, who would regard these beliefs as being unsupported by credible empirical evidence. On the other hand, believers are likely to regard such attitudes as evidence of the closed-mindedness of most mainstream scientists. For now we shall let the pejorative "weird" label stand, and defer to the next chapter a discussion of one particular paranormal area.

It seems likely that the pronounced rises in belief in some of the listed items are probably due as much to the nature of "documentary" television shows that explore them than to a greater degree of public gullibility. For example, at one time such documentaries would present a relatively balanced look at paranormal phenomena, including both skeptics and believers, but

Table 4.1

Percentages of Americans Who Express Various Beliefs

Belief	2001	1990	1978
Psychic/spiritual healing	54	46	—
Extrasensory perception	50	49	51
Possession by devil	41	49	39
Ghosts	38	25	11
Clairvoyance	32	26	24
Astrology	28	25	29
Witches	26	14	10

87   that is rarely done today. In fact, the rise in certain beliefs can be directly linked to specific "documentaries." In July 1999, for example, only 11 percent of Americans believed that the U.S. lunar landing was just a hoax, but that percentage doubled following two televised showings of "Conspiracy Theory: Did We Land on the Moon?"[30]

The important role of the media in contributing to the rise in superstitious belief in the United States is explored in the book *How Superstition Won and Science Lost*, by John Burnham.[31] Burnham argues convincingly that the diffusion of scientific knowledge in society has over a long period of time proceeded in three stages that have led to a progressive deterioration, and that we are now in the third stage:

- **Stage 1: Popularization**—scientists themselves attempt to share their ideas with the public.
- **Stage 2: Dilution**—popularization is done primarily by journalists and educators, rather than scientists.
- **Stage 3: Trivialization**—the science being conveyed consists mainly of isolated snippets, and tends to emphasize exciting "Gee Whiz!" science, that often tends to reinforce mystical beliefs rather than promote a reasoned viewpoint.

Burnham's critique is probably correct regarding science as most often presented in the nonprint mass media, but it must also be said that many fine popular science books continue to be written by first-class scientists. An interesting aspect of the so-called weird beliefs is the correlation between the percentage of people who believe them and the believer's level of education. For example, consider table 4.2, which lists beliefs and percentages of believers, based on the 2001 Gallup data for four educational levels. The direction of the arrows on the right indicates whether the belief tends to be held more widely by those with less (←) or more (→) education. It is quite interesting that some of these beliefs appear to be held more widely by the more educated, while others are more widely held by the less educated. My use of the pejorative term "weird" to refer to these beliefs may be more appropriate for some of them than

Table 4.2

Percentage of Persons Expressing a Belief in Various Phenomena as a Function of Their Educational Level

Belief	Percentage Believers by Educational Level				
	1	2	3	4	
Psychic/spiritual healing	48	57	57	65	→
Telepathy	35	36	38	41	→
Extrasensory perception	49	48	53	52	
Possession by devil	46	41	35	32	←
Haunted houses	48	41	37	33	←
Astrology	33	29	20	16	←
Witches exist	24	30	24	22	

Note. 1 = high school or less, 2 = some college, 3 = college graduate, 4 = postgraduate. The direction of the arrows on the right indicates whether the belief tends to be held more widely by those with less (←) or more (→) education.

others. One particularly tricky item is psychic or spiritual healing, defined in the Gallup Poll as "the power of the human mind to heal the body." Unfortunately, this definition could include anything from "psychic surgery" (shown to be complete quackery) to the well-known ability of mental stress to have an effect on your immune system and hence your recovery from an illness (see chapter 8).

Another reason for not attributing too much significance to a rise in "weird" beliefs over the last decade is that each generation seems to be susceptible to its own particular brand of such beliefs. Consider, for example a pair of polls conducted in 1925 and 1950 on strange beliefs prevalent in 1925 (table 4.3).[32]

The six listed beliefs were from a much longer list of 31 items. It is interesting that belief in the six items (and many of the other 31) declined dramatically between 1925 and 1950. Also, with the exception of the astrology item, none of the other 31 items have appeared on the Gallup Polls conducted since the 1970s, which probably indicates that few people any longer be-

89 **Table 4.3**

Changes in Strange Beliefs between 1925 and 1950 Based on Polls of
College Students

	% Believing	
Statement	*1925*	*1950*
Long, slender hands show an artistic nature.	42.0	6.4
Adults can become feeble-minded from overstudy.	56.0	10.9
You can closely judge a person's IQ from his face.	50.0	3.6
Women are by nature purer and better than men.	38.0	1.8
The position of the planets affects your character.	15.0	6.4
Expectant mothers can affect the character of unborn children by thought.	38.0	2.7

lieve this particular collection of nonsense. It is noteworthy, however, that belief in astrology has increased dramatically since 1950.

Are Stupid People Outbreeding Smart Ones?

Perhaps the most controversial kind of evidence that people are getting dumber (and, of course, I mean those *other* people, not you or I) lies in the realm of "dysgenics." Francis Galton formulated the concepts of dysgenics, and its opposite eugenics, in the latter part of the nineteenth century.[33] Both concepts are based on two assumptions: (1) intelligence is inherited to some degree, and (2) people of higher intelligence have fewer children than average. It then follows that the next generation will tend to be less intelligent than the present one.

These ideas, which were championed by a number of thinkers in the first half of the twentieth century have become less respectable since then, largely because of the uses to which they have been put. For example, during the 30-year period before 1937 some 32 states in the United States passed laws that

90   allowed forced sterilization of those deemed undesirable. This category included the mentally ill, the handicapped, and persons convicted of sexual or drug-related crimes. Eugenics (and its modern descendant, the idea of genetically engineering "designer babies") therefore has not-so-faint echoes of a desire to create a "master race," which reached its apex (or nadir?) in Nazi Germany. Similarly, dysgenics—the idea that those *other* intellectually inferior people are outbreeding the rest of us— has obvious connotations of racism. Nevertheless, these bad associations of the concept of dysgenics should not keep us from taking an objective look at it, and evaluating whether there in fact might be a dysgenic trend in human intelligence.

The two assumptions on which dysgenics rests are that intelligence is inherited to some degree, and that less intelligent people tend to have more children. Nearly all psychologists agree with the first proposition, although the extent of heritability of intelligence is open to question. Many estimates based on studies of twins and adopted children suggest that anything from 40 to 80 percent of your intelligence is inherited (genetic), with the rest determined by your environment, or the interaction of the environment and your genes.[34]

The other assumption about less intelligent people tending to have more children is more difficult to establish. If we accept that IQ tests are a measure of intelligence, however, a number of studies tend to show an inverse correlation between IQ and fertility, i.e., higher IQ means lower fertility (table 4.4).[35] (A positive correlation means the relationship is eugenic, and a negative correlation means that it is dysgenic.)

The extent of the correlation seen in the eight studies differs, which could indicate some variation with time, with the early years indicating a eugenic trend (positive correlation) and the later years a dysgenic one (negative correlation). However, a more plausible explanation may lie in the limited and unrepresentative samples of the first three studies (showing a eugenic relation). Those three studies are less reliable than the five later ones, because they relied primarily on urban or largely urban white populations. Had the early studies included rural popu-

Table 4.4

Correlation between Intelligence and Fertility in the United States

Study Authors	Year	Correlation
Bajema1	1963	+.05
Bajema2	1968	+.04
Waller	1971	+.11
Osborne	1975	−.49
VanCourt	1985	−.16
VanCourt	1985	−.29
Vining	1995	−.06
Vining	1995	−.23

Note. Studies in this table are identified in reference in note 35.

lations (which tend to have higher fertility), they may well have found a negative correlation. In sum, the U.S. studies slightly favor a dysgenic trend. Other European studies conducted since 1954 have produced mixed results, but with the dysgenic findings outnumbering the eugenic ones four to one.

Even if we were to accept both of the ideas behind the concept of dysgenics, i.e., heritability of intelligence and greater fertility of the less intelligent, it still doesn't necessarily follow that society will actually get progressively dumber over time. The less intelligent tend to be economically poorer and less aware of proper nutrition and healthy living; therefore, they face greater health problems that affect their chances of survival until their reproductive years. Others have raised additional questions about the reality of a dysgenic trend in intelligence.[36]

Moreover, even if there is a genuine dysgenic trend taking place, its actual magnitude is rather small. Based on the value of the correlation between IQ and fertility, some estimates suggest that IQ might drop by only about a point from one generation to the next.[37] Of course, a few points' loss in IQ generation after generation, and before too many centuries, we'd be a bunch of blooming idiots—*assuming* the theory is right. The main problem with the dysgenic theory is that no such decline

92   in IQ is seen. In fact, quite the reverse appears to be occurring: IQs are steadily rising![38] (A steadily rising average IQ doesn't mean that dysgenics is nonsense, only that the modest decline in intelligence it predicts is not observed, possibly because other offsetting factors are more important.)

Are People Getting Smarter?

The strange phenomenon of rising IQ scores has been called the "Flynn effect," after New Zealand philosopher James Flynn.[39] Isolated data supporting the Flynn effect were around for many years, but Flynn is credited with the observation, because he noticed the broad pattern that had escaped general notice. When IQ tests are periodically updated, some people are given both the old version and the new version. Flynn found that if one test is 30 years older than the other, then on the average, the same people taking both versions score about 10 IQ points higher on the old version. In other words, someone scoring at

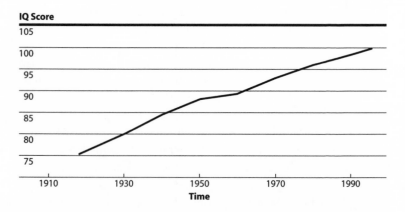

Figure 4.1 IQ scores versus time. IQ scores of white Americans from James Flynn.[38] Mean scores have been normalized to 100 in 1990. Reprinted with permission from J. Horgan, Get smart, take a test. *Scientific American* 273(5):14 (1995). Copyright © 1995 Scientific American.

93   the mean of today's IQ test would score well above the mean when compared to people who took the test a generation ago, so the Flynn effect cannot be simply attributed to IQ tests becoming easier. As figure 4.1 shows, the result is a nearly constant rise in IQ score over time extending back almost a century.[40]

Data are now available for 20 nations in the developed world, and all show a rise, but by varying amounts. (The increases were highest in Belgium, Holland, and Israel at 20 points per generation, and only half that in Denmark and Sweden.) When Flynn made this strange discovery he suspected that perhaps the explanation was that people were simply becoming more familiar with the specific kinds of information contained on such tests. To test this idea he looked at the results of scores on the so-called Ravens matrices tests over time (see a sample test item in figure 4.2).[41] Rather than testing

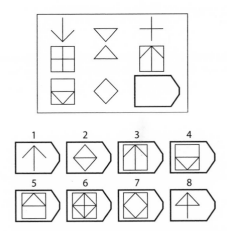

Figure 4.2 A problem similar to those in the Raven Advanced Progressive Matrices. Find which of the eight numbered items best fits into the lower right similarly shaped element in the top portion of the figure. From P. A. Carpenter, M. A. Just, and P. Shell, What one intelligence test measures. *Psychological Review* 97:409 (1990). Copyright © 1990 American Psychological Association, reprinted with permission.

94   general knowledge, the Ravens test supposedly measures what psychologists call "fluid intelligence," or the ability to formulate and test hypotheses without relying on general knowledge.

If Flynn were right about the source of rising IQ scores, he expected to see little or no rise in Ravens scores over time. To his great surprise, Flynn found that the Ravens scores were rising even more rapidly than the IQ scores, and the result again held for many nations (figure 4.3).[42] The important point about figure 4.3 is that the slopes of the lines are roughly similar for each nation; the different heights of the curves are not meaningful. The increases Flynn found in the Ravens test scores were so large that someone who scored in the top 10 percent a century ago would score in the lowest 5 percent today.

There are only two explanations for the Flynn effect: either people are getting more intelligent (for some reason or other) or they are not and the Flynn effect is due to some kind of artifact. As we shall see, each type of explanation has problems. The main difficulty in giving a good explanation of the Flynn effect is that the rise in IQ that it describes is both so huge and fairly

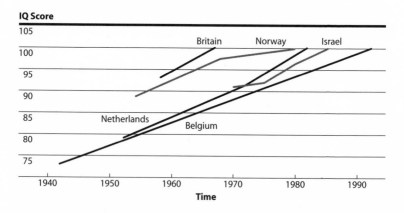

Figure 4.3 IQ scores versus time. Comparison in IQ gains in five nations, based on the Ravens test: Britain, the Netherlands, Israel, Norway, and Belgium (from Flynn[38]). Because of the way the scores are normed, only the slopes are meaningful in making comparisons between nations, not the average heights. Copyright © 1998 American Psychological Association, reprinted with permission.

95 uniform (within a factor of two) across time and across the various nations and age groups examined.

Suppose the Flynn effect is due to some kind of artifact. A likely possibility might be that people are simply getting better at taking these kinds of tests due to greater familiarity, being that students now are bombarded with standardized tests in school. One problem with that idea is that the IQ gains preceded the period of standardized test taking, and have continued rising at a uniform rate, even after people became test savvy. Also, the Ravens test is not one that most people are generally exposed to during their school years. Another possible artifact might be biased sampling, i.e., that smarter groups of people tend to take the tests as time goes on. But that idea cannot explain the Ravens data, since for the countries in which the test is used to test military recruits, virtually all young men of military age are tested. Nor does the biased sampling argument hold for the IQ test data, because the same people taking new and old IQ tests tend to do better on the old versions, even when the tests are taken in a random order.[43]

If the Flynn effect does represent a real rise in intelligence over time, the rise would certainly seem to be due to environmental rather than genetic factors. Evolution takes many generations to yield major changes in a population and could not possibly work on such a short timescale for humans, where a generation is roughly 30 years. Many possible environmental changes might account for a rise in intelligence over time, including

- More years of schooling
- More technology in school and work
- More exercise of visual skills
- Better nutrition
- More urbanization
- More technical jobs

Years of schooling, for example, could account for a real rise in intelligence over time all on its own, depending on how you define intelligence. (It is conceivable, for example, that people are getting dumber, but that the decline is more than offset by

96 increased years of schooling.) Each additional year of schooling accounts for more than one year of age in a child's performance on IQ tests.[44] During the past century a phenomenal rise has occurred in the average number of years of schooling in many developed countries, especially in the United States. Unfortunately, this explanation has a major flaw: the rise seen on IQ tests appears smallest on school-related subjects, such as vocabulary and arithmetic, and it is largest on questions relying on basic reasoning skills, such as deducing similarities. So, the Flynn effect could represent a real rise in intelligence. Flynn himself, however, points out that if the rise in intelligence is genuine, a significant puzzle remains—were our ancestors really idiots by our standards, or are we really geniuses by theirs? Neither possibility seems plausible.

Given an IQ rise of as much as 10 to 20 points in one generation, your parents' generation would have to be classified as nearly mentally retarded by your generations' standard. While that possibility might seem plausible to some teenagers, it probably doesn't fit many people's everyday observations. For example, do most teachers toward the end of a 30-year career really have the experience of finding themselves surrounded by mostly near-genius students compared to their own generation? (Many American teachers might come to the opposite conclusion.)

One group of educational researchers has sought to answer the preceding question empirically by surveying older teachers in three nations: Australia, Korea, and Singapore.[45] Their results for the two Asian nations, which have experienced rapid environmental improvements in recent decades, suggest that children really are getting brighter. They blame the negative result for Australian children on a possible decline in motivation that could offset a rise in real intelligence.

Are There Real-World Indicators of Rising Intelligence?

If the Flynn effect really does indicate that people are getting smarter, we would expect to see evidence of this gain in some real-world accomplishments beyond the classroom. Patents

97   were considered earlier as one possible indicator, but as already noted, little can be concluded based on variations in the numbers of patents issued. One other realm that has been suggested as a good place to look for possible evidence for a rise in intelligence over time is the game of chess. Chess has a number of advantages as a test of the hypothesis that real-world intelligence is rising:

- It has objective measures of level of skill.
- Its rules have not changed for a long time.
- Skill in chess correlates closely with IQ.
- There are no barriers to entry into the game.
- Statistics on competitors are tracked.
- All ages compete against each other.

One recent paper by Robert Howard has claimed that the statistics for world-class chess players do indeed offer preliminary evidence for the notion that real-world intelligence is climbing.[46] Howard's claim is based primarily on the observation that the fraction of younger players among those in the top 50 in the world seems to be rising (figure 4.4). Unfortunately, this evidence seems to be rather shaky for a number of reasons.

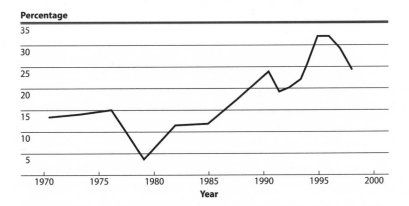

Figure 4.4 Chessmasters under 25. Percentage of chess players among the top 50 players who are under age 25. Reprinted with permission from an article in *Intelligence* by R. W. Howard.[20] Copyright © 1999 Elsevier Science.

98   First, with only 50 top players, the fraction under 25 years of age is subject to very large statistical fluctuations. Second, there are many explanations why chess might catch on among young people at a given time that have little to do with their ability, and as the overall pool of young players expands, we would expect to find a greater the number of highly talented ones. Third, the graph in figure 4.4 seems to be consistent with the fraction of expert players under 25 years old starting to rise only around 1986, which is not consistent with a uniform rise in IQ lasting over the last century. In fact, if the rise in intelligence really were occurring at a roughly constant rate over a very long time, we would predict that the fraction of young players among the top 50 should be constant rather than rising, as each new generation of upstarts easily vanquishes the previous generation.

Another study purporting to show a possible real-world increase in human mental ability over time considered the realm of politics. The authors of that 1997 study claimed that elected officials, the press, and people called to give Congressional testimony have on the average all shown an increasing level of skill over the course of the twentieth century.[47] Unlike chess, however, skill in politics is much more in the eye of the beholder. The skill levels of the individuals being rated in this study by Rosenau and Fagen were based on their written statements. In the appendix of their study, Rosenau and Fagen list their criteria for defining eight skill levels. Their highest level of "integrative complexity" is one for which

> The speaker constructs and presents a "global view" of "multiple controversies," acknowledging trade-offs among conflicting goals in his explanation of "complex interactions" within a complex system.

In contrast, the authors' lowest-rated skill level is one in which events are interpreted or choices made with little or no room for ambiguity. On such a rating scheme, we might suppose that politicians and journalists who take unambiguous positions on issues or had to take uncompromising positions would have fared badly. Alternatively, political "leaders"

99   might score highly if they recognize all the various complexities of obscure issues of no great consequence, or seek to find justification for inaction. Clearly, a different sort of rating scheme might result in precisely the opposite conclusion of Rosenau and Fagan, so that evidence of increasing levels of intelligence or skill based on such an analysis is highly suspect.

Another Look at Standardized Test Trends

The performance of American students on standardized tests, while it may not demonstrate declining intelligence or ability, doesn't show evidence for the converse either. Lacking any hard evidence that the real-world intelligence of people is increasing over time, what are we to make of the Flynn effect of rising IQ scores? Someone who is critical of the U.S. educational system could argue that people really are getting smarter (as evidenced by rising IQ scores), but that an increasingly poor educational system cancels out the effect so as to produce students whose standardized test scores are essentially constant over time.

Aside from the strange coincidence of having a rising IQ that just offsets a supposedly deteriorating educational system, the magnitude of the IQ rise seems too great and too prolonged for such an explanation to be credible. Furthermore, essentially constant SAT scores (since the 1970s decline) actually may be consistent with steadily rising achievement levels, because the fraction of 18-year-olds who take this test continues to rise—up from 40 to 59 percent over the last 25 years. In addition, the 0.4 percent of students exceeding 1500 on their SATs has quadrupled since 1976 when only 0.1 percent achieved that score.[48]

A Possible Explanation of the Flynn Effect?

James Flynn himself appears to have changed his mind about the meaning of the effect named after him. For some years

100   Flynn was of the opinion that the long-term rise in IQ was not associated with a rise in real-world mental ability, and remained uncertain as to what might account for it. (One reason for Flynn's skepticism was that if changing environmental conditions had to explain a 10-point rise in average IQ from one generation to the next, those changes would apparently need to be implausibly large.)

More recently, Flynn has teamed up with William Dickens of the Brookings Institution and proposed a specific mathematical model to account for the rise in IQ scores.[49] In the model Flynn and Dickens suggest that the rise in IQ scores really does imply rising cognitive abilities over time. Flynn and Dickens suggest that a feedback mechanism exists in which positive external and internal changes create still more positive changes that enhance an individual's mental ability, so that each success breeds further success, as a person's confidence and opportunities grow. The particular societal changes that drive these changes and account for the rise in IQ vary with time and location. Moreover, the overall rise in society's average cognitive ability feeds back to create a new challenge to individuals, causing them to increase further their own performance.

Such a positive feedback effect is clear in other areas, such as sports, where as the overall level of play increases over time, individual players step up to the challenge. The gradual increase in performance is most obvious in individual sports, such as swimming and track and field, where new records continue to be set. But it also occurs in team sports, where the result of such increasing skill levels sometimes yields paradoxical results. Thus, the number of 300-hitters in baseball today is far less than in the past, because while batters are hitting better than ever, their improvement is more than offset by the improvements in pitching and fielding.[50] A similar effect is likely to occur in more cognitively demanding areas for which success depends on competition, making it difficult to observe a rise in skill level.

Most intellectual pursuits do depend on competition to some degree. But, even when competition is not an important factor, the need to surpass (and build on) the achievements of the past

101 makes increases in skill level difficult to ascertain. To make new discoveries in science, for example, researchers have much more powerful tools than scientists in the past, but they also must do much more sophisticated and complex work, since "the easy stuff" has already been done. Consider this comparison between a pair of classic physics experiments over four centuries apart.

Galileo, a famous sixteenth-century Italian physicist, wanted to observe the motion of falling objects. Lacking any way to precisely time a falling object, Galileo realized he could in effect "dilute" gravity by rolling objects down an inclined plane. The resulting motion was then slow enough that he could time the ball's descent using his own pulse and a water clock. Galileo's experiment is now done and understood by many students in high school and college physics classes. Nowadays, instead of rolling a ball down an incline, students observe a cart as it moves along a low-friction "air track" using electronic detectors (not water clocks!) to measure the time (figure 4.5).

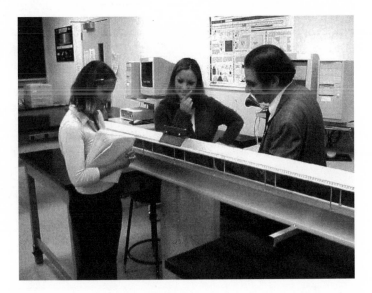

Figure 4.5 Two George Mason University physics students discussing a version of Galileo's experiment with their professor.

102   By way of contrast to Galileo's classic sixteenth-century experiment, consider the recent "neutrino oscillation" experiment done by an international Japanese-led team of over 100 physicists. This experiment used a detector containing 50,000 tons of ultra-pure water and state-of-the-art devices built specifically for the experiment that recorded the data over a period of many years. The inside of this detector when it was half-filled with water is shown in figure 4.6. The experimenters recorded reactions initiated by subatomic particles known as neutrinos, which are found to change their "flavor" (oscillate) between their point of creation in the sun and their detection in the apparatus. The results and significance of the neutrino oscillation experiment are probably not understood by most physics Ph.D.'s. Yet, I imagine that most of those physicists involved in the neutrino oscillation experiment would not claim that their

Figure 4.6 The Super-Kamiokande neutrino detector while it was being filled with water. Notice the small boat on the right side of the photo for a sense of scale. Picture courtesy of ICRR (Institute for Cosmic Ray Research), The University of Tokyo, printed with permission.

103 own mental prowess exceeds that of the sixteenth-century scientist with his simple experiment with rolling balls.

As this comparison shows, a steady rise in mental sophistication and complexity can occur, which is made possible by more complex tools, greater competition, more schooling, and, most importantly, the accumulated wisdom of the past. This increase in sophistication over time can possibly account for the Flynn effect—though it is open to question whether that change represents a rise in intelligence. (I'd guess that were Galileo to time travel to the present, he would be able to become a very productive modern physicist after some catch-up period. Of course, no empirical test of my suggestion is yet possible unless some clever physicist builds a time machine—see Epilogue.)

My rating for the idea that people are getting dumber over time is 2 flakes. I give only 1 flake to the idea that they are getting smarter, because it seems a bit more plausible, given the Flynn effect.

5 Can We Influence Matter by Thought Alone?

The force is the energy field created by all living things;

it binds the galaxy together.

—*Obi-Wan Kenobi of* Star Wars

PROFESSOR K is a good friend and colleague who thinks that my mainstream scientific view of the universe prevents me from understanding some of its more mysterious and intriguing aspects. Although K is a professor in the humanities, he has followed with great interest some of the new discoveries in quantum theory, which some physicists believe shed light on the nature of human consciousness—perhaps the greatest mystery of all. My colleague continues to have hope that I will eventually see the light, and while we continue to disagree about almost everything, we have great fun discussing the implications of modern physics for consciousness and other areas, such as alternative medicine.

Professor K knows that I am an empiricist, who is willing to believe in the reality of phenomena, even if I lack a theory to explain them, and possibly even if they contradict our current understanding of the laws of nature. Accordingly, he recently made me an offer I couldn't refuse: he offered to pay for me to go see a therapeutic touch practitioner, so that I could become convinced for myself that there was something to this particular branch of alternative medicine. Therapeutic touch is considered by some observers to be a form of "energy healing," and is one form of spiritual or psychic healing. The therapist does not actually make contact with the patient in therapeutic touch. Instead, by moving her hands near the body she can supposedly sense the presence of a person's "aura" or energy field that surrounds the body, and thereby diagnose problems. (Touch

105 therapists disagree over just how the healing process works, with some believing they transmit healing energy—typically by pointing one or two fingers at various places—and others believing the process involves a mind-to-mind connection.[1])

I did get a slightly funny look from the touch therapist when in our initial interview I was unable to articulate any particular problem that brought me to her for treatment. (I didn't feel comfortable telling her about the exact circumstances that brought me to her office, and mumbled something about "curiosity.") During the actual therapeutic touch session she was able to determine that the energy level of my "core chakra" was dangerously low, and she then proceeded to "balance" my chakras—at least that's what I thought she said. This diagnosis was apparently made in part by moving her hands near my body, and in part by using a pendulum held over my body.

My friend, Professor K, later demonstrated for me the utility of a pendulum for observing the energy field around one's body. While holding a beautiful crystal pendulum near his own chest he observed how vigorously it started to swing spontaneously, as compared to when it was near my chest— remember my low core chakra "problem." As a physicist familiar with the phenomenon of resonance, I explained to my colleague that what he was observing was not the response of the pendulum to energy fields surrounding our respective bodies, but rather subtle (unconscious?) motions of his hand holding the pendulum.

Psychic Pendulums—An Amusing Demonstration

The following amusing demonstration has been excerpted from my book *Why Toast Lands Jelly-Side Down.*[2] It requires a pair of pendulums hanging side by side from a bar held in the hands. The pendulums might consist of two filled cans of soda with labels attached to them reading "yes" and "no" (figure 5.1). The strings holding the two pendulums should differ in length by at least 20 percent. When one of the cans is driven reso-

106 nantly by shaking the bar very slightly (without observers being able to perceive the motion) at just the right frequency, the hanging cans seem to answer yes/no questions asked of them by the audience. With a little practice, you should find that it is quite easy to excite only one can into large oscillations by moving the bar so imperceptibly that the can seems to start moving spontaneously. For best control of the motion of the bar you may wish to sit down and hold the bar in two hands while resting your elbows on your knees. When you get proficient at exciting one can by tiny motions of the bar, you almost may convince yourself that it is responding to your psychic powers!

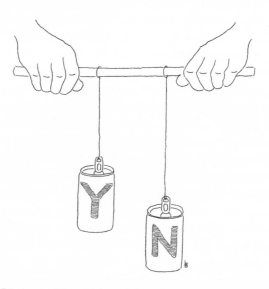

Figure 5.1 A pair of "psychic" pendulums for answering yes/no questions. From Ehrlich, *Why Toast Lands Jelly-Side Down.* Printed with permission from Princeton University Press.

The idea of resonance is that very tiny amplitude vibrations fed into a system can cause very large vibrations if the frequency of the outside energy source matches the "natural" frequency of the system. (A famous example of this phenomenon

107 is when an opera singer causes a wineglass to shatter if she hits just the right note.) My colleague Professor K at first refused to believe that the motion of his hand (rather than the energy field of his body) was actually the source of the pendulum's movement.

Only when I suggested setting up an experiment where the pendulum was suspended from a fixed support above his body did he concede that it was his hand that was responsible for the pendulum's movement after all. None of this, however, in any way diminished his belief in the energy fields surrounding our bodies. Perhaps my colleague believed that the therapist's mind "resonates" with the energy field of her patient, and this in turn is what causes her hand holding the pendulum to move unconsciously.

Are there energy fields surrounding our bodies? Of course there are. Weak electromagnetic signals can be picked up by electrodes placed very near the heart, eyes, and brain. Such a detection shows that parts of the body do produce weak conventional electromagnetic signals or energy fields detectable outside the body using suitable apparatus—but certainly not with something as crude as a pendulum.

If the human brain does generate a weak electromagnetic field that can be detected in the space around it, you might wonder whether it is also able to produce other fields that have much greater influence, perhaps even some all-encompassing "unified field." The so-called unified field of mainstream physics actually reduces to separate known fields, including gravity and electromagnetism, at any temperature we are ever likely to achieve. The unified field described by physicists is certainly not some mystical all-encompassing field, which includes human consciousness, as some have claimed. (It is conceivable that other energy fields exist that science has not discovered, but if they interact with the human brain, one would imagine they also should be measurable by instruments.)

The idea that mind can influence matter has long been part of popular folklore. Gamblers who try to influence mentally the roll of a pair of dice or a ball in roulette cannot be very success-

108 ful over the long term, however, given the small advantage in odds that the house is willing to be satisfied with, and still make a large profit. In other words, if mind can influence matter, for most people the influence would have to be a very small one, otherwise the house would go bankrupt.

On the other hand, even a small effect could allow you to make a fortune if you knew when your psychic powers were "on." You could simply wait until those propitious times to place your bets, and consistently win. Of course, the idea that you could know when your powers were "on" might be wrong if these powers operated at an unconscious level. Apart from such financial considerations, even the slightest effect of mind over matter would have enormous significance, both philosophically and practically, if we could only find a way to amplify it.

Some might argue that this debate is moot, because we already have the technology that accomplishes such amplification. For example, at least one company, Brain Actuated Technologies Inc, manufactures a computer interface that allows paralyzed persons to play video games and generally interact with a computer.[3] The interface works by picking up small electrical signals the paralyzed person is taught to generate and control by using biofeedback.

In one demonstration of the technology, a company spokesman was said to have steered a boat supposedly by thought alone. However, despite the company's claim, and indeed its name, it seems likely that the generated electrical signals from brain waves are far too small to have been observed, and that the observed electrical signals were instead produced by the movement of facial muscles or eyeballs. There are, however, more relevant experiments done at Brown University in which a paralyzed man was able to control a cursor based solely on his thoughts, and the associated signals generated by particular groups of neurons in his brain. It is hoped that scientists will be able to transform these lab experiments into useful applications for the disabled allowing them to control some apparatus. But, such important technologies are somewhat off the point of this

109 chapter. They say nothing about whether our minds can directly influence distant matter in the absence of electrodes that detect and amplify brain waves.

Possible Relevance of Quantum Theory

Most mainstream physicists today scoff at the notion that mind can influence external matter. That was not always the case. In fact, many of the leaders in the development of quantum theory, including Bohr, Heisenberg, and Pauli, have speculated about the relationship between physical reality and consciousness.[4] As Bohr noted, "The real problem is how can those parts of reality that begin with consciousness be combined with those parts that are treated in physics and chemistry?"[5]

In quantum mechanics it is assumed that a system such as an atom need not be in a single well-defined state at any given time. Rather, the system can be described as being in a combination or superposition of states described by its so-called wavefunction. Some physicists today argue that the "Copenhagen interpretation" of quantum mechanics lends support to the idea that mind can influence matter. In the Copenhagen interpretation, various possible states of a system are present simultaneously until a measurement is made, at which point the system's wavefunction "collapses" suddenly to a single state. The famous example is that of Schrodinger's fictitious cat. Imagine an unfortunate feline that is sealed in a box with a vial of cyanide that breaks when a random event occurs. According to quantum theory, we must consider the cat to be in a kind of limbo state that is simultaneously alive and dead until the box is opened and the "measurement" of its state is made.

A less dramatic example of the collapse of the wavefunction would involve the position of an electron in a beam of electrons that falls on a pair of slits. The position of any one electron in the beam is not defined, but rather it follows a certain probability distribution determined by the wavefunction. But after an

110 electron's position is measured, the probability distribution collapses to a single point, where that electron is actually found. Perhaps, say some physicists, it is the mind of the observer that guides the wavefunction collapse to that particular point.

Those who make such arguments also sometimes cite the supposedly nonlocal character of certain quantum mechanical interactions involving "entangled" pairs of particles.[6] In one version of an experiment proposed by Einstein, Podolsky, and Rosen (EPR), a measurement of the spin direction of one electron determines instantly the spin direction of the other, even though the two electrons may be separated by a vast distance. Extending this result to include the realm of consciousness, one might speculate that the observer's mind and the external universe also form a single entangled system in which parts can affect one another even when they are separated in space and time through nonlocal interactions.[7]

The nonlocality of physical interactions is only one possible interpretation of the EPR experiment. Consider the following analogy to that famous experiment. Instead of a pair of electrons, suppose that we prepared an initial state consisting of one man and one woman, both initially located in a tent. The man and the woman leave the tent simultaneously and travel in opposite directions, while covered head to toe in garb disguising their gender.

When one member of the pair reaches a certain point and a "detector" reveals his/her sex, that of the other member of the pair is also simultaneously revealed, without any need for a second "measurement." Nonlocal interactions were unnecessary to draw this conclusion, because the initial "sex-neutral" state ensured the result. Exactly the same thing can be said of the actual EPR experiment, where the man corresponds to a "spin-down" electron and a woman a "spin-up" electron, and the lack of spin in the initial state corresponds to sex neutrality.

The EPR experiment is somewhat more complex than the man—woman experiment, owing to the rules of quantum mechanics. To make the two situations parallel, we would need to find that once a party's sex was detected, and he/she again

111 donned a head-to-toe covering garment, a second measurement of some different property such as hair color placed the
individual in a limbo state regarding sex. In other words, once
in the limbo state, another sex determination could with 50 percent probability lead to a spontaneous sex change! To summarize, what is really illustrated in EPR-type experiments is not a
refutation of locality, but rather a refutation of "reality" in
quantum mechanics, i.e., the idea that it is possible to know
with 100 percent certainty a property of a system without the
need for a measurement.

Are Paranormal Phenomena Consistent with Physical Theory and Practice?

Many mainstream scientists, such as Nobel Laureate physicist
Philip Anderson, have been critical of paranormal claims and
speculations, because they argue that were the claims valid,
they would violate many fundamental laws of physics.[8] In particular, if it really were true that mind could influence matter,
and do so without regard to distance or time as some have
claimed, we would have to throw out the such basic ideas as
causality, i.e., the idea that cause must precede effect.

In this chapter we want to closely examine the evidence for a
particular paranormal ability, namely the ability of mind to influence external matter or information, which is usually called
psychokinesis or PK. (Sometimes the related term telekinesis is
used, but that refers to the supposed ability to make objects
move, whereas PK allows for other types of influence aside
from movement.) Before even considering the empirical evidence, however, we might ask whether PK could be consistent
with the laws and practices of physics. Skeptic Anderson has
expressed the view that precision measurement as routinely
practiced in physics would be impossible if PK were real.[9]
However, Anderson's view may be overly conservative. It
seems possible that a sufficiently small PK effect could be
unnoticed or hidden in the noise of many measurements.

112 Moreover, the phenomenon of experimentalists finding values for a quantity that converges on some theoretical prediction, even as the theoretically predicted value changes systematically over time is well known to many physicists. Could that effect be an example of PK?

Conventional wisdom explains the type of convergence referred to above as being due to a natural human tendency of experimentalists to find "corrections" to their data when they disagree with a theory, but not to look at the data as closely when they agree. An alternative but much more far-fetched explanation would be that the mind of the experimentalist literally affects his measurements, making them conform better to theoretical predictions, whenever those predictions change.

Thus, based on this example, we cannot be certain that PK is incompatible with our ability to make precise measurements, as Anderson has claimed. Ultimately, whether PK really exists is an empirical question. Mainstream scientists are correct to demand extraordinarily strong evidence in support of extraordinary claims, as the late astronomer Carl Sagan used to say, but they should not dismiss empirical claims that happen to be in conflict with the laws of physics, *as we now know them.*

It has happened many times in the past that phenomena were dismissed out of hand as being impossible, but later shown to be real. We must be on guard that we don't treat scientific laws as if they are religious dogma that forbid certain events from occurring. Large stones actually do fall from the sky, ball lightning does exist, and the continents do drift, even though scientists initially believed these events to be impossible. In fact, Carl Sagan himself, while remaining extremely skeptical about all "new age" beliefs, noted that several paranormal claims, including PK, deserved further serious study.[10]

Claimed Observations of Psychokinesis

PK is one of a number of so-called paranormal phenomena that are claimed to exist. A working definition of the paranormal

113 might include those phenomena that are not presently accepted by most scientists. Aside from PK, they include telepathy (mind reading), precognition (seeing future events), and clairvoyance (detecting images and other impressions of a distant scene), also called remote viewing.

Initially, reports of these alleged phenomena were associated with specific "psychically gifted" individuals. Experiments to observe such phenomena in the lab began in the nineteenth century, when telepathy was first studied. PK was first looked at in the lab in the experiments by J. B. Rhine and colleagues at Duke University in the 1930s.[11] Most researchers doing this work now claim that such abilities are present in all people to varying degrees.

Invariably, however, when the "gifted" individuals are tested under extremely rigorous controlled conditions, their psychic abilities seem to disappear. Moreover, psychic debunkers, such as magician James Randi, have shown specifically how tricks can be done, when conditions are not rigorously controlled. In fact, Randi's foundation has a standing offer of a one-million-dollar prize to "anyone who can show, under proper observing conditions, evidence of any paranormal, supernatural, or occult power or event."[12]

Some believers in the reality of paranormal powers claim that Randi has "rigged the game," however, by demanding particular things that he knows they cannot accomplish, for example, reading numbers off a dollar bill hidden in an opaque envelope. People said to be capable of "remote viewing" claim to experience shapes and other aspects of a distant scene, but they cannot read numbers. Some believers in the paranormal also claim that the powers of a "gifted" individual can melt away due to bad vibes or a "shyness" effect when confronted by a skeptical audience who insists on rigorous controls, but skeptics are likely to come to another conclusion.

Surprisingly, many people's belief in paranormal abilities is so strong that even when they witness how psychic abilities can be duplicated using trickery, their belief does not waver. Perhaps they reason that while some individuals may be resort-

114 ing to trickery, surely there must be a core of genuine psychics, and even "genuine" psychics may on occasion need to enhance their performance through tricks. Many believers, including my good friend Professor K, also have the sense that debunkers, such as the Amazing Randi, are not to be applauded for their efforts in exposing frauds. Rather, they view such debunkers as never being satisfied that enough controls are in place, and being so closed-minded that they could never accept real evidence for paranormal phenomena, even after witnessing the phenomena themselves.

I think, however, the preceding views misread the skepticism that all scientifically minded persons should bring to any revolutionary claims. Most scientists are not experts at the art of deception, so it would be difficult to rule out fraud unless airtight controls are in place. Insisting on such airtight controls is reasonable when we are evaluating a claim that if true would turn our present understanding of the universe on its head.

Nevertheless, even if all "psychically gifted" individuals are frauds or self-delusional, it is still just possible that PK and other paranormal abilities are present in all of us to a small degree. Experimental PK research has been conducted over the years by many individuals in a number of laboratories, some of which have an affiliation with prestigious universities. The research is usually published in journals dedicated to paranormal or "anomalous" phenomena, rather than in mainstream scientific journals. Should that fact raise our suspicions? Not necessarily. Articles submitted to mainstream scientific journals are likely to be reviewed by mainstream scientists before being accepted. Most scientists asked to review an article that claimed evidence for paranormal phenomena would (justifiably) start out from a position of extreme skepticism. It is, therefore, understandable that researchers in this field would believe they are not getting a fair hearing, causing them to start their own journals.

What do typical psychokinesis experiments look like? Most of them no longer involve attempts to move stationary objects, such as bending a spoon, or causing an object to levitate or move across a table. Such claimed results have been shown to

115 be due to trickery in the past, and they are something of an embarrassment to the field. On the other hand, some psychic researchers do claim to have witnessed the movement of objects in the presence of specific persons ("recurrent spontaneous PK" or "poltergeists"), under circumstances that they claim could not be faked. William Roll, one such psychic researcher from the University of West Georgia, helpfully explains that such spontaneous activity of objects can result from their becoming "emotionally charged," so that they "no longer are constrained by inertia and gravity after they absorb zero point energy from the vacuum."[13] Unfortunately, "explanations" of this sort involve so many new claims that they are meaningful only if we don't probe too deeply into what is being said.

Before considering what modern PK experiments look like, a further word about levitation might be in order. Various means exist to levitate objects, such as using magnets (which have even been used to levitate trains), but we are here, of course, considering the idea of causing objects to levitate by our mental powers alone. We are so used to seeing magicians levitate their assistants on stage that some viewers may be forgiven if they sometimes forget that such levitation is strictly an illusion done using various tricks.

As far as levitation done using only our mental concentration, some people claim to be able to achieve this feat through meditation in "yogic flying," where people are able to jump into the air from a cross-legged lotus position. Although most people may not be able to do this (or even sit in a lotus position to start with!), it seems strange to regard yogic flying as levitation, since it could be simply the result of a well-developed gluteous maximus. Real levitation would involve staying in the air for a while after leaving the ground, something no one has been able to accomplish without the aid of technology. There are groups who dispute this claim, stating that historically documented cases of real levitation exist (see www.yogicflying. org), but the "evidence" presented consists only of anecdotes.

Most modern PK experiments involve trying to alter random processes by willful intention. A random process results in some kind of statistical distribution, which someone tries to al-

116 ter by thinking about it. Of course, distributions do have statistical fluctuations, so researchers need to be sure that if they do see a departure from chance, it actually is statistically significant. There are several alternative explanations why claimed evidence for PK now shows up only in such random processes, and not in the movement of stationary objects. One possibility is that nonlinear chaotic processes can be extremely sensitive to external perturbations, which make them a good way to look for a small effect. (The oft-cited example is of a butterfly flapping its wings somewhere, which leads to a chain of events culminating in a storm appearing elsewhere.) Moreover, while it used to be thought that random noise invariably degrades measurement, recent research shows that adding some noise can actually enhance our ability to detect faint signals.[14] A more cynical explanation why PK experiments now use statistical distributions of random processes is that it is much easier to inflate claims using statistics, particularly when the object under study is a human ability about which the researcher may have a strong belief.

Since the early work of J. B. Rhine there have been many hundreds of experiments studying PK, or in more general terms the "anomalous interactions of human consciousness with physical systems." (Some researchers prefer this latter term, because it is more encompassing, but for brevity we'll just use PK throughout.) In the remainder of this chapter we shall analyze the results claimed by these experiments and of one particular series of experiments done by Robert Jahn and his colleagues at Princeton University. Jahn's work is generally considered to be among the highest quality in the field, because of its technical sophistication and the sheer volume of data he has collected.

Research by Robert Jahn and His Colleagues

Robert Jahn was the Dean of the School of Engineering and Applied Science at Princeton University when he established

117 his Princeton Engineering Anomalies Research (PEAR) laboratory with his associate Brenda Dunne (the lab manager) in
1979. Jahn acknowledges he is viewed as a pariah by many of
his mainstream colleagues on campus. (Depending on your
view of Jahn's research and the purpose of tenure, this example
shows either the merits or the flaws of the tenure system in
place at most American universities. However, even those aca-
demics who are skeptical of the research might praise a system
that provides safeguards for those who study very controversial matters.)

In the 22+ years of its existence, the activities of the PEAR
lab have centered on experiments to study "anomalous
mind–matter interactions." The results of these experiments
have been disseminated in over 53 articles and technical reports listed on the PEAR web site (www.princeton.edu/~
pear), and some of them can be directly downloaded. Most of
the published articles have appeared in the *Journal of Scientific
Exploration*, whose editorial board includes Robert Jahn.[15]

This journal, despite its generic name, generally publishes articles that deal with a variety of paranormal phenomena, including dowsing, UFOs, PK, and remote viewing. While the
journal does publish some skeptical and critical articles and re-
tains some skeptics on its board, many of its articles are written
from a point of view that admits the reality of paranormal phenomena. In other words, the journal is at the other end of the
spectrum from *The Skeptical Inquirer* and *Skeptic*, two journals
published by and for "debunkers" of such claims.

Dr. Jahn's research generally has involved experiments in
which someone (an "operator") tries to alter the outcome of a
random distribution of events or numbers by thought alone.
The effect being observed is not on the events individually, but
rather on their observed distribution. For example, if the random events were coin flips, the operator doesn't try to influence whether any one flip comes up heads or tails, but rather he
tries to make more heads or tails turn up after some large number of flips. (It could, of course, also be argued that the effect actually operates on individual coin flips, but only once every so

118 often is the size of the operator's mental influence great enough to alter the outcome.)

In most of Jahn's experiments the events are truly random, rather than "pseudorandom"—which would be the case for the deterministic series of numbers that a computer can generate, since the sequence is reproducible. A variety of different means have been used to generate the random events, including mechanical, optical, fluid flow, and electronic devices. For example, in one device resembling a pachinko game, thousands of

Figure 5.2 A random mechanical cascade device. Courtesy of Princeton Engineering Anomalies Research (PEAR) Laboratory.

119 balls fall through an array of pins, and form a bell-shaped distribution in bins located at the bottom of the device, while an
operator seated about six feet away tries to shift the distribution to one side or the other by sheer thought. (This particular
device, shown in figure 5.2, is no longer in active use because of
the noise it makes and the length of time it takes for all the balls
to land in the bins.)

One of the most beautiful and ingenious devices built at the
PEAR lab is a crystal pendulum that swings on a low-friction
bearing. The pendulum is in a glass case, but small random air
currents can affect the speed of its swings. When these small
variations occur, the glowing crystal is made to change colors—
red for slightly faster than normal, blue for slightly slower. An
operator viewing the pendulum attempts to affect its speed by
thought alone, i.e., make the glowing crystal stay red or blue.
(In this case, unlike Professor K's hand-held crystal pendulum,
an effect cannot be the result of hand motion.)

Another PEAR lab device is their "robot frog" (figure 5.3).
The robot is powered by a circuit that causes it to follow a series

Figure 5.3 A robot frog device on a table. Courtesy of the PEAR Laboratory.

120 of steps constituting a random walk. If the frog is placed initially at the center of a circular table, it has a 50/50 chance of reaching the edge of the table on one side of an arbitrary center line. An operator tries to wish for the frog to reach his side of the center line more often than the opposite side. This particular device is understandably very popular with visiting groups of children, who are said to be more successful than adults in mentally affecting froggy's random path.

One device the PEAR lab has built, which they call a random event generator or REG, is used in the large majority of their experiments—see Jahn's publications for details on the construction and operation of this device.[16] The REG uses thermal "white" noise that is converted to a series of bits (zeroes and ones) that are produced in "trials" of 200 bits at a time. As one guard against machine bias, every other bit in the series is flipped when it is generated.

On the average, it would be expected that the results of any one trial would include roughly 100 ones out of 200 bits, subject to the usual statistical fluctuations. It is very much like flipping a coin 200 times and keeping track of the number of heads, but the REG data can be generated much more rapidly than data from flipping a coin. The PEAR lab has taken great care to ensure that the data REGs produce are truly random by having a "failsafe" equipment design, operating procedures, and most importantly through periodic calibration runs. In such calibration runs the number of ones per trial is observed to closely follow the predicted bell-shaped Gaussian distribution with the expected mean of 100.

In most of the experiments an operator sits in front of the REG for about an hour (the time needed to generate 1000 trials). Operators are said to have no physical contact with the device, but since they are not monitored continuously, one cannot be certain of this claim.[17] (Operators are free even to conduct their trials with the door shut if they wish not to be disturbed.)

The operator's instructions are simply to think about and wish for the REG to generate one of three types of sequences of trials: high, low, and baseline. For high (or low) runs, the oper-

121 ator wishes for more (or less) than 100 ones per trial on average, and for "baseline" runs he has no preference. In some cases, the operator chooses which of the three cases to wish for, but usually the choice is predetermined by a random choice made by the computer. Whatever the operator's intention (high, low, or baseline), this preference is recorded in the database manager *before* the REG is activated and data taken.

In the absence of any effect of human intention on the bit distributions, we would expect that within statistical fluctuations the high, low, and baseline distributions should be indistinguishable. During a run the operator usually sees the results displayed on the face of the REG and also on a computer screen, so he can tell how well he is "succeeding" in his attempt to influence the result. (Quotes are used around succeeding, because statistical fluctuations could be interpreted falsely as success or lack of it.)

At the time of this writing, the PEAR lab has run over 14 million trials during a period lasting more than 12 years, using over 108 different operators.[18] The operators are unpaid anonymous volunteers, none of whom claim any special "psychic" abilities. The results from these experiments have been reported in a number of articles and technical reports, where it is claimed that statistically significant effects of operator intention are observed. Although the magnitude of the claimed effect is only several extra or fewer ones for every 10,000 bits, the effect is said to be statistically significant with a probability of being due to chance of only one in three trillion, or just over seven standard deviations for the entire data set.[19]

Standard Deviations and Measurement Uncertainty

The standard deviation is a measure of the width of a distribution of events or measurements. For a Gaussian or bell-shaped distribution, 68 percent of its total area falls within the region one standard deviation on either side of the central peak, and 95 percent of the area falls within two

122 standard deviations. This means that in experiments in which the measurements often follow a Gaussian distribution the chances of a single measurement falling no further than one (or two) standard deviation(s) from the true value are 68 percent (or 95 percent), respectively. Conversely, if we report a measured length (with a one standard deviation uncertainty) as, for example, 6.3 ± 0.1 meters, we imply that the true length has a 68 percent chance of being in the interval 6.2 to 6.4 meters. By chance it would be almost unheard of to find a measurement as far as seven standard deviations from the expected value. Therefore, when we find deviations this large the implication is that something other than chance is the explanation.

Although Jahn and Dunne acknowledge that some operators produce stronger results than others, it is claimed that there are no "superstars" responsible for the bulk of the effect. Even more amazingly, statistically significant results are claimed to occur regardless of the spatial or temporal separation between the operator and the REG, even if the operator makes his wish thousands of miles away, or *before or after the run is made*!

In fact, it is claimed that within statistical fluctuations, the size of the effect (several bits per 10,000 above chance) does not depend on either distance or time separation.[20] If true, these results would be inconsistent with the expected decline in signal strength with distance and, more significantly, with the concept of causality, i.e., that a cause (an operator's intentions) must precede the effect (the change in the bit distribution). These are truly extraordinary claims, which, in Carl Sagan's words, require extraordinary evidence, so let's take a close look at the evidence compiled by the PEAR lab and other researchers.

A Critical Look at Some PEAR Lab Results

The results for which the PEAR lab has reported on in the greatest detail are those obtained using their electronic random

123 event generator (REG). An exploratory set of runs consisting of 5000 trials were conducted with one anonymous operator (identified only as "operator number 10"). These results are depicted in figure 5.4, which shows a plot of the number of bits excess over what is expected by chance versus the cumulative number of trials conducted.[21]

Think of this plot as the number of excess heads (over the expected 100) found versus the total number of sets of 200 coin flips. However, the curves show *cumulative* results as more and more trials are conducted. The three raggedy looking curves correspond to the three intentions of the operator (high, low, and baseline [no operator preference]). For a given observed excess (or deficiency), the number of standard deviations for a given number of trials grows as the square root of the number of trials. As a result, the curve corresponding to a particular number of standard deviation excess or deficiency is a parabola. The two parabolas straddling the zero line correspond to plus or minus 1.645 standard deviations, or a probability against chance of 5 percent ($p = 0.05$). (These parabolas are shown for reference, because in many fields of research departures from chance are considered statistically significant when they reach this level.)

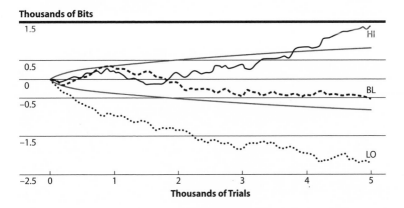

Figure 5.4 Cumulative deviations from mean for high, low, and baseline trials. First 5000 trials for operator 10. Reprinted with permission from Dunne and Jahn.[20]

124   The curves depicted in figure 5.4 would seem to indicate a highly significant effect, since they wander well outside the two parabolas. When the operator wishes for fewer ones (the lower "LO" curve), the cumulative total after 5000 trials is well below the zero line by a statistically significant amount (with a probability of being due to chance of less than one in 100,000 or $p = 0.000009$). The opposite is true when the operator wishes for more ones, although the observed deviation is not quite as impressive, and only breaks through the $p = 0.05$ parabola after about 4000 trials. The baseline trial results are quite consistent with chance.

Intrigued and encouraged by these early results, Jahn was able to gain additional funding allowing him to embark on a more robust program extending over many years. For example, when data were aggregated from this same operator (number 10) over a 12-year period (120,000 trials), Jahn and Dunne continued to find statistically significant deviations correlated with the operator's intentions (figure 5.5).[22] The HI discrepancy now shows the greater departure from chance, whereas in the initial run it was the LO, but perhaps the best measure of significance would be the HI minus LO departures, which is highly signifi-

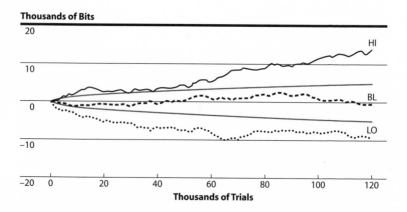

Figure 5.5 Cumulative deviations from mean. Twelve years of data for operator number 10. Reprinted with permission from Dunne and Jahn.[20]

125 cant in both figures 5.4 and 5.5. These are labeled "local" data,
since they don't include other trials in which the operator tried
to influence the REG from a remote location, possibly thousands of miles away.

To investigate whether similar effects could be achieved by
other operators, Jahn and Dunne compiled local data taken
with 91 different operators over a 12-year period (figure 5.6).[23]
Again, note that the LO and particularly the HI results show
significant departures from chance in the direction of the operator's intention.

Having described the REG results, let's now cast a critical eye
at them. Have you perhaps noticed anything unusual about the
data presented so far in figures 5.4–5.6? Normally, when progressively more data are accumulated, the statistical significance of a result increases, assuming the initially reported result is a genuine effect, rather than a statistical fluctuation or a
mistake. Here, however, just the opposite result is found—the
statistical significance gets weaker as more data are added.
Even worse, as we shall see, the extremely strong results found
for operator 10 are by themselves responsible for much of the
departure from chance seen for all 91 operators collectively.

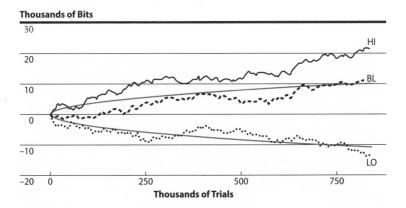

Figure 5.6 Cumulative deviations from mean. Twelve years of data for all operators.
Reprinted with permission from Dunne and Jahn.[20]

126   Let us now take a close look at the data in these figures to see
how well they support the claim of a genuine anomaly. The values in table 5.1 have been extracted from figures 5.4–5.6 by
noticing where each curve is at the end of the highest number
of trials shown. For example, to obtain the first row of entries
that refer to the first 5000 trials for operator number 10, we see
from figure 5.4 that the curve labeled HI is at about 1350, and
the LO curve is at about −2150 at the right end of the graph.
The overall statistical significance depends on the observed difference between the HI and LO values in column 3 of the table.
Thus, the "HI minus LO" entry is 3.5 thousand in this case. The
one standard deviation uncertainty in that value, here ±0.7
thousand, is given by $10\sqrt{N}$, where N is the number of thousands of trials. Finally, the same two quantities are shown in the
last column after dividing each by N.

The values listed in the last column represent the *effect size*.
For example, for all 91 operators combined (third row of table)
we find an effect size of 0.042, meaning that there are on the
average 0.042 extra ones or zeroes per 200 bits. This is the
equivalent of one extra head in roughly 4800 coin flips—an exceedingly small, but supposedly statistically significant result,
given the small uncertainty of 0.011. *The claim of a genuine anomaly here depends on whether the effect size listed in the last column is
significantly greater than zero, assuming that any systematic error or
cheating can be ruled out.*

Table 5.1
Statistical Significance of Data in Figures 5.4–5.6

Number of Trials	Operator	HI minus LO	
		In 1000s	*Per 1000 trials*
5,000	#10	3.5 ± 0.7	0.70 ± 0.14
120,000	#10	19 ± 3.5	0.16 ± 0.03
850,000	All	35 ± 9.2	0.042 ± 0.011
730,000	All but #10	15 ± 8.5	0.021 ± 0.012

127   Consider the following observations about the data tabulated in the last column.

1. *Consistency?* One operator (number 10) initially showed a "signal" 33 times stronger than all other 90 operators tested later, i.e., 0.70 is 33 times 0.021, so that the effect size seen in the early PEAR results are completely inconsistent with the later ones. This inconsistency is downplayed in PEAR publications, which emphasize probabilities, i.e., p values over effect sizes.

2. *Significance of overall result?* With the data for operator 10 removed from the total data set (last row of table), the overall effect departs from chance by only 1.8 standard deviations.

3. *No superstars?* Although the effect without including operator number 10 is only marginally statistically significant, Jahn and Dunne repeatedly have claimed that there are no "superstar" operators, and that "several have produced larger absolute effect sizes" than operator number 10.[24] What they mean by this statement is that some operators have values of the HI minus LO difference (the "absolute effect size" shown in the last column) that are further than zero than those of operator number 10. Of course, most of these other operators also have much less data and much larger uncertainties in their results, so it is not surprising that their "effect size" might be larger, strictly on the basis of chance. (Taken alone, the data set for operator 10 is 5.6 standard deviations from chance, while only one other operator exceeded three standard deviations.[25])

4. *Who is operator number 10?* Jahn and Dunne have chosen to preserve the anonymity of all the operators, including number 10, on the grounds that they are considered as collaborators rather than research subjects. However, the following known facts about this operator raise some concerns. First, he or she has been providing data for a period of over 12 years, and has been with the lab since the very first pilot study. He or she has also accumulated roughly 15 times as much data as the average

128 operator. The longevity of this operator and his/her being the first one ever tested suggests that he or she is not a casual drop-in student. Moreover, according to a conversation with Dr. Dunne, both she and Dr. Jahn have been operators, and the three most prolific operators over the years have all been female. These facts suggest that operator 10 is Dunne herself, or perhaps one of Jahn or Dunne's female associates. When explicitly asked in person if she could rule herself out as being operator number 10 Dunne refused to do so in order to "preserve operator anonymity." Dunne's refusal can be considered an indirect admission that she is operator number 10, because the anonymity of operator 10's identity would not have been compromised by a simple denial.

If operator number 10 were either Dunne or one of Jahn or Dunne's associates, this person would probably have enough familiarity with lab procedures to bypass the usual precautions that would prevent cheating by a casual operator. This is *not* to accuse operator 10 of cheating, but only to point out that she would apparently have that capability. One simple form of cheating would be to change the classification of a trial after it was conducted. For example, if a baseline trial turned out to yield more (or less) than 100 ones, a cheater could record it as a HI (or LO) trial, so as to give apparent agreement with the stated intention. This form of cheating would result in the baseline trials being much more narrowly bunched (have a smaller standard deviation) than chance would predict.

In fact, this type of pattern is precisely what is seen in the data, as Jahn and Dunne candidly admit. Making the situation even stranger, when the abnormally narrow baseline distribution is added to the other data, the combined distribution is perfectly bell-shaped (Gaussian) and has just the right width, giving further reason to suspect that some baseline trials have been accidentally or deliberately misclassified as HI or LO trials.

Of course, Jahn and Dunne do not raise the issue of such a suspicious pattern in their baseline data, because it suggests possible cheating. Rather, they suggest that such a pattern

129 could be relevant to a theoretical model of the phenomenon. For example, given the *assumption* that they are observing a real phenomenon, such a pattern in the baseline data could be quite understandable. As an operator observes the cumulative graph being generated on a computer screen during a baseline run, he might turn on his "psychic power" to drive the curve back toward zero if the baseline graph started wandering too far above or below the zero line.

When asked about operator cheating, a member of the PEAR staff assured me that they deliberately try to cheat before putting protocols in place formally. However, the decisions to not monitor operators, to allow them to work in a closed room, and to work on a computer where the records are stored leads to questions about how cheat-proof the protocols can be. For example, the data file for a given trial is only about 50 characters long. Given a knowledge of file format and naming conventions, an operator could change the trial type (HI, LO, BL) in a matter of seconds from the computer keyboard.

In fairness to the PEAR lab, while their protocols may not be cheat-proof, cheating of a magnitude necessary to explain their results might not be an easy undertaking. The preceding suggested scenario of categorizing baseline trials after the fact as HI or LO was specifically discussed with the staff member in charge of security and Dr. Jahn. According to Dr. Jahn the only person with enough technical expertise to accomplish this feat would be their head of security. In particular, operator number 10 was claimed to lack the necessary expertise to cheat in this manner.

The "Most Frequently Asked Question"

Robert Jahn notes that the question most frequently asked about their results is why they don't try to amplify the effect they are seeing by taking data at a much higher rate. In fact, they actually have taken some data at two million bits per trial instead of the typical 200—a ten thousand times

130 higher rate. If their effect were real, and were not dependent on the data rate, we might expect a statistical significance that scaled with the square root of the number of bits, or a gain of a factor of 100 when the number of bits per trial increases 10,000-fold.

In fact, however, they see nothing of the sort. When runs are made using two million bits per trial instead of 200, the overall effect size is only just under four standard deviations, and it is reversed in sign! In other words, departures from chance are found in the opposite direction from the operator's intentions: operators wishing for HI runs tend to get LO ones, and vice versa. If one believed the results were genuine, one might argue that, as with any human ability, psychic abilities cannot be expected to be independent of the rate at which we process data, and that the mind loses the ability to influence external data above some rate. But, then why the reversal in sign?

A further reason for skepticism about the PEAR results is supplied by their attempt to replicate these high data rate trials. According to PEAR researcher York Dobyns, "The replication experiment generated results . . . quantitatively somewhat different from the original. The scale of the difference is sufficient to make the question of 'successful replication' somewhat ambiguous."[26] The inability to amplify alleged paranormal phenomena, or even replicate them reliably, is perhaps their most troubling feature, causing most scientists to question their existence.

The PEAR lab has published other data aside from the local REG results involving individual operators. These include data where an operator is remote from the REG, in which my suggested scenario of operator cheating is even more problematic.[27] A number of the PEAR studies also looked into the effect of an emerging "collective consciousness" in influencing and being influenced by external events. (This work is described at their "global consciousness web site" noosphere.princeton. edu/. This site describes, among other things, how major world events, including the September 11, 2001 terrorist attacks supposedly were detected by their impact on the collective data from many people's REG readings around the world.

131   One of the "group consciousness" studies considered whether the local Princeton weather could be influenced by people's wishes for good weather. The study claims to have observed a statistically significant correlation between special occasions held on the Princeton campus and the frequency of good weather on those occasions (compared to that in surrounding areas).[28] This particular finding is an interesting illustration of the perils of making an "informed choice" when testing a hypothesis for statistical significance, which was discussed in chapter 2.

The idea goes something like this. Given a normal statistical variation, it is clear that in some places the weather will be better than average on certain specific days, and in other places it will be worse. The Princeton weather study was done precisely because it was often remarked that special events on the Princeton campus in recent years were almost always blessed by good weather. Thus, to select a place (Princeton) where conventional wisdom says that the weather has been almost always good on certain days is to make an *informed choice*. In this case, a finding that indeed the weather was in fact usually good on those specific days proves nothing.

The proper test of the hypothesis would be to ignore the weather data from the past, and see if *future* days when special events are held on the Princeton campus are blessed by better than average weather. (The fallacy in the PEAR analysis of Princeton weather is very similar to what we explored in figure 2.6. In that case, recall that we noted that any random distribution of dots is bound to show some clustering. After one finds a cluster, i.e., an above average density of points, it is not proper to find the probability of the higher than expected density of points without taking into account all the other places we might have looked.

However, without looking at any more of the PEAR claims, given the earlier detailed analysis of the results with the REG device, it is clear that the evidence that they present falls far short of being "extraordinary." (I should again stress that nothing said here should be taken to imply that Jahn or Dunne have

132 cheated—indeed, both researchers seem to be quite sincere and dedicated individuals—only that on the basis of the evidence presented, cheating cannot be ruled out with a high degree of confidence.)

Other Relevant Experiments

Since Rhine's early experiments there have been over 800 separate studies of PK using REG-type devices. When faced with a large number of scientific experiments all supposedly measuring the same phenomenon, a standard technique is to combine the results in a so-called meta-analysis. For example, given many separate measurements of the speed of light, c, or the mass of the electron, m, we can combine them to find the best values of c and m, by taking a weighted average of the individual measurements, with weighting factors depending on the uncertainties, so as to favor the more precise measurements.

However, the kind of meta-analysis usually done by PK researchers is of a very different sort, since the very existence of the phenomena is at issue. By combining the results of many experiments, none of which is conclusive by itself, researchers hope to find an overall result that is definitive. This notion that many weakly positive results can combine to give a significant one is questionable. It is generally not followed in the hard sciences, except to find the best value of a quantity. For example, in particle physics, the results of many experiments might be combined to yield the best value of the mass of the electron, but many weakly positive experiments would *never* be combined to establish that electrons exist.

Apparently, however, this practice is followed in medicine and in other areas, including education research. For example, if many medical studies individually show a benefit from some new drug, the totality of the studies in a meta-analysis can yield a statistically significant result, even if the individual studies lack such significance. Medical studies, however, have the advantage of usually being double-blind, which is not the

133 case for experiments testing the reality of PK or other paranormal phenomena. They are not double-blind because both the operator and possibly the experimenter know when the operator is trying to influence the distribution of bits to be higher or lower than average.

As we have noted, many weakly positive experiments are a poor substitute for a single convincing one—just as many weak arguments cannot substitute for one convincing one—and that is precisely what is lacking in paranormal research. But why, you may wonder, is the combination of many statistically weak separate results any different than a single experimenter combining many data runs (which are individually statistically insignificant) to yield an overall result that can be highly significant? The difference is that the individual researcher is using a common methodology throughout the experiment. It is much more perilous to combine many separate experiments involving different methodologies and flaws. Aside from the problem that some of the experimenters may have inflated their results either through poor controls or outright mistakes, there is also the "file drawer" problem, where some experimenters who obtain negative results simply choose not to publish them.

When combining data from many experiments the overall result will have a much smaller statistical uncertainty, but the systematic errors cannot be assumed to "average out." To appreciate the distinction between statistical and systematic errors, just imagine averaging many individual measurements of the diameter of a sphere made by people who don't understand parallax, and make the measurement by always putting a ruler in front of the sphere, rather than behind it. The average of those many individual measurements will have a small statistical uncertainty, but it will be smaller than the sphere's true diameter because of parallax—go ahead and try it! In this case, a careful experimenter would either use a different measurement technique or else correct for this source of systematic error. But, in practice, small sources of systematic error become increasingly important (and harder to detect), the smaller the statistical uncertainty becomes when many experiments are combined.

134   Putting these concerns aside for the moment, let's take a look
at a meta-analysis conducted by Dean Radin and Roger Nelson
of over 800 experiments. These authors claim that their analysis
shows that the overall PK effect is around 16 standard deviations from chance—an extraordinarily significant result, if correct. They obtained this seemingly impressive result by combining the z-scores (the number of standard deviations) of large
numbers of experiments, few of which on their own gave positive results in excess of two standard deviations. In fact, as to be
expected, a number of the individual experiments actually
gave *negative* z-scores, or less of a PK effect than would be predicted just on the basis of chance.

Another unsatisfactory aspect of the meta-analysis is how
the results have changed over time. Figure 5.7 shows the cumulative overall z-score versus the year of publication of the experimental results beginning in 1959.[29] Notice how there is a
rapid rise in the cumulative number of standard deviations up
to nearly 14 in 1965, after which the number stays roughly constant. Given that more than 90 percent of the experiments
reported are 1965 or later, this leveling off of the cumulative
number of standard deviations is extremely peculiar. What it

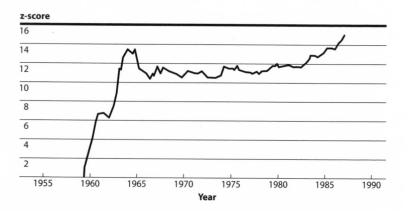

Figure 5.7 Cumulative significance. Cumulative z-score through 1987. Reprinted
with permission from Radin and Nelson.[29]

Figure 5.8 "Darn, I meant to think one lump, not two!"

implies is that the small number of early experiments gave nearly all the effect observed, and the bulk of the experiments conducted post-1965 gave almost none of it.

Obviously, people's PK powers did not suddenly turn off in 1965, so we are led to conclude that controls were not as tight in the earlier experiments as in the later ones, causing the earlier experiments to give a spurious positive result. Even though Radin and Nelson believe their analysis does indicate a highly significant positive result, they do acknowledge that the earlier experiments were lacking in many controls, which steadily improved over time.

For example, one simple technique for getting a spurious positive result in some early experiments was to allow subjects to begin and end trials as they wanted. That way, if the result seemed favorable in a given trial they would continue to completion and record the result. Conversely, if a trial was going

136 badly, they could abort it either on the grounds of a headache or an assessment that their psychic powers seemed not to be working that particular day.

One further aspect of Radin and Nelson's meta-analysis that casts doubt on their claim of the reality of PK is their finding that for the 500+ experiments studied there was a slightly *negative* correlation between the number of events in an experiment and the reported statistical significance.[30] Normally, it would be expected that the statistical significance of a result would grow with the number of bits or events studied. For example, if you possessed the genuine PK power to create 7 extra heads for every 200 coin flips, you would produce a one standard deviation excess for 200 flips. But, after repeating the exercise for a total of 800 flips (four times as many), your excess of 28 extra heads would represent two standard deviations, and after 3200 flips, you would reach four standard deviations, with the statistical significance growing the more data you accumulated.

In fact, as Radin and Nelson note, the actual experiments do not show this pattern, but remain no more statistically significant (actually slightly less so!), even as their number of events increase significantly. Their conclusion, based on this peculiar observation, is that PK does not operate on individual random events, and that the effects being observed "cannot be explained by simple, linear, force-like mechanisms."[31] Whatever that explanation may mean, presumably the claim is that one's PK power somehow shifts the observed statistical distributions without affecting individual events. However, even if one accepts this strange claim, it is not self-consistent, because when an experiment looks at distributions compounded from many individuals, the result should be more statistically significant than with few.

In an attempt to explain why the phenomena under study do not show the usual statistical behavior, Radin suggests that "in research that touches the boundaries of our understanding, we must remain open to less canonical possibilities."[32] But that explanation appears to suggest that embracing the reality of para-

137 normal phenomena may require that we modify the laws of statistics as well as those of conventional physics. In summary, the overall evidence for the reality of PK based on Radin and Nelson's meta-analysis seems extremely weak.

In evaluating the evidence that mind can have a direct effect on external matter or statistical distributions of external events, I would give the idea a 4-flake rating (very flaky).

6 Should You Worry about Global Warming?

Predictions are difficult, especially about the future.

—*Niels Bohr*

MANY PEOPLE believe that global warming could have grave consequences for life on Earth. Research on the subject is regularly monitored and disseminated by the Intergovernmental Panel on Climate Change (IPCC), a respected international group of hundreds of scientists. According to a recent IPCC report issued in 2000, portions of which are downloadable from the IPCC web site at www.ipcc.ch, the rate of global warming projected for the coming century "is very likely to be without precedent during at least the last 10,000 years."[1] Yet, in seeming contradiction to that alarming statement, MIT meteorologist Richard Lindzen, one of the lead authors of the very same IPCC report, notes that "the possibility of large warming, while not disproven, is also without a scientific basis."[2]

Lindzen, a member of the prestigious National Academy of Sciences, is said to be "the nation's most prominent and vocal scientist in doubting whether human activities pose any threat at all to climate."[3] The nature of the global warming issue is such that it is unclear even how to refer to the differing positions. Nearly all sides agree that (1) *some* amount of global warming is likely in the coming century, and (2) the amount might not be very large. The main fault line is probably between those who believe that mitigation measures should or should not be taken *now* to avoid the effects of future climate

139 change that might be significant. We'll refer to the latter group of people as global warming skeptics or, for shorthand, skeptics.[4]

Be Forewarned

This chapter is considerably longer and more detailed than many of the others. It also has many more qualifiers of the "on the one hand, on the other hand" type. Global warming is a complex subject, and if we want to do it justice, we need to consider the details that are often left out of popularized accounts of the subject. If you want to jump ahead and get a sense of my bottom-line assessment of the problem, see my answers to four commonly asked questions on page 182.

Where does public opinion stand on the issue? Roughly two-thirds of Americans surveyed in a series of Gallup Polls have indicated that they worry either a "fair amount" or a "great deal" about the problem, with roughly equal percentages for the two levels of concern.[5] Those percentages have changed little over the last decade. One reason for caution in assessing public opinion is that while people claim to worry about global warming, an April 2001 poll found that the issue ranked next to last out of thirteen environmental concerns, ahead of only acid rain (table 6.1).

While you're contemplating where you fit on the global warming opinion spectrum, join me on an imaginary raft trip. The following beautiful allegory opens George Philander's book *Is the Temperature Climbing?*[6]:

> We are in a raft, gliding down a river, toward a waterfall. We have a map but are uncertain of our location and hence are uncertain of the distance to the waterfall. Some of us are getting nervous and wish to land immediately; others insist

Table 6.1

Percentages of Americans Who Worry "a Great Deal" about Various Environmental Problems,* Listed in Descending Order of Concern

Environmental Problem	Americans Who Worry (%)
Pollution of drinking water	64
Pollution of rivers, lakes, and reservoirs	58
Contamination of soil and water by toxic waste	58
Contamination of soil and water by radioactivity	49
Air pollution	48
The loss of natural habitat for wildlife	48
Damage to the Earth's ozone layer	47
The loss of tropical rain forests	44
Ocean and beach pollution	43
Extinction of plant and animal species	43
Urban sprawl and loss of open spaces	35
The "greenhouse effect" or global warming	33
Acid rain	28

* According to a March 2001 Gallup poll.

that we can continue safely for several more hours. A few are enjoying the ride so much that they deny that there is any imminent danger although the map clearly shows a waterfall. A debate ensues but even though the accelerating currents make it increasingly difficult to land safely, we fail to agree on an appropriate time to leave the river. How do we avoid a disaster?

Philander argues that his allegory of the drifting raft captures our dilemma about global warming. The connection becomes clearer when we consider the causes of global warming. Among its other causes, global warming arises from increases in the concentration of so-called greenhouse gases (such as CO_2 and methane) in the atmosphere. Global warming clearly has both natural and human-caused or anthropogenic components. The main anthropogenic components are due to the emission of

141 CO_2 from the combustion of fossil fuels, and the clearing and burning of forests.

Given that the world's fossil fuel usage continues to increase, and that atmospheric CO_2 levels have been rising (figure 6.1), it can only be a matter of time before dangerous levels are reached, assuming that growth continues unabated. People may disagree about what level is dangerous and when it will be reached, but the nature of unlimited growth is such that sooner or later we will face a disaster. Furthermore, the longer we wait to address the problem, the more difficult will be the solution.

Even worse, it could be a long time between the implementation of mitigating measures and any impact on climate, because it takes a considerable length of time for elevated levels of atmospheric CO_2 to return to equilibrium due to natural removal processes.[7] At some point, we may find that we are no longer able to prevent the global warming equivalent of having our drifting raft swept over the falls.

The growth in human population is only partly responsible for the growth in atmospheric carbon dioxide level. It is reasonable to expect that standards of living will rise in the developing nations, where the current per capita energy usage is a mere

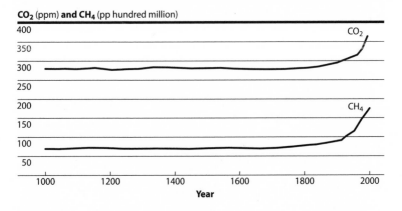

CO₂ (ppm) and CH₄ (pp hundred million)

Figure 6.1 Growth of atmospheric carbon dioxide (upper curve) and methane (lower curve) as a function of time. CO_2 concentrations are in parts per million and methane (CH_4) concentrations are in parts per hundred million.

142 11 percent of that in the developed world. Thus, it is possible that the future growth in atmospheric CO_2 levels may even be faster than that of population due to a growth in per capita energy usage. The possibility of continued growth in CO_2 levels seems to place the notion of an impending climate catastrophe in the category of a logical syllogism:

1. Atmospheric greenhouse gases have been increasing unabated.
2. Anything that increases unabated will eventually reach a catastrophic level.
3. Catastrophe cannot be avoided unless steps are taken to slow the growth.

In other words, there surely is a big waterfall somewhere on the river up ahead of us, and the only question is how soon would it be prudent to take our raft out of the river, i.e., reduce greenhouse gas emissions, so as to avoid disaster. However, things are not quite so simple. Most predictions of future global warming assume that growth in human population will end in a few decades. Were we to assume otherwise, no remedial action would be possible, and disaster would await us from any number of other causes, aside from global warming. Once we assume that global population will stabilize, however, we no longer can be certain that greenhouse gases will continue to rise unabated and that disaster necessarily lies ahead. As a result, global warming skeptics could argue that the allegory of the raft drifting toward a catastrophic falls needs to be modified so as to capture all the subtleties of the global warming problem.

Possible Modifications to the Allegory of the Drifting Raft

The relevance of each modification to global warming is shown in italics.

1. We don't know for sure if the waterfall actually exists or whether the map actually applies to this particular section of the river. In fact, the people who drew the map have not visited this section of the river, but instead used computer models to infer the existence of a large

143   waterfall. *(Aside from the evidence provided by computer models, we really don't know for certain whether global warming will be a very serious problem in the future.)*

2. If there is a waterfall, we'll hear the sound of the falls from such a great distance that we can delay our exit from the water until then. Of course, this would be a safe course of action only if the river current does not accelerate very quickly until we are within hearing distance of the falls. *(Given the long time span involved, action on global warming can be put off until its seriousness becomes clearer.)*

3. Even if the waterfall does exist, it is likely to be sufficiently small, so that a trip over it would actually be fun. Obviously, we would not make such an assumption unless we had some good evidence that the falls are very likely to be small—though members of our party might disagree as to how large a falls is navigable. *(The benefits of global warming might actually outweigh the harm.)*

4. All the waterfalls on this river are believed to have a chain barrier in front of them that prevents boats from getting too close. *(Nature will prevent a catastrophic warming without human intervention being required through the action of negative feedbacks that reduce the amount of warmings.)*

5. There are dangers on shore—poison ivy or snakes, perhaps—so that we want to delay our exit from the water as long as possible. Possibly, the dangers on shore are so great that they are even worse than going over the falls. *(The cost of simply adapting to global warming, i.e., "going over the falls," might be less than the cost of preventing it.)*

These five modifications to our raft scenario would likely induce a variety of reactions in us as we drift down the river in our raft: watchful waiting (1, 2), complete lack of concern (4), resignation (5), or eager anticipation (3). We might be engaged in an animated debate among ourselves depending on who tended to believe in one or the other modified scenarios. All the while, our raft continues to head downstream toward . . .? The preceding more complex allegory is possibly closer to "the boat we're in" than Philander's original version, when it comes to the global warming debate.

144   Those who believe we shouldn't be too concerned about global warming base their belief on a variety of arguments. Some skeptics claim that there is no evidence any warming has taken place, even though theoretically it should have been observed by now if the models are correct. Other skeptics will concede that some warming has taken place, but there is little evidence it was human-induced (or anthropogenic) In addition, most of the skeptics dispute the predictions of future warming, arguing that there is little evidence it will be nearly as dire as some researchers have predicted. As a final line of defense, some skeptics argue that even if global warming occurs in the future, it is likely to prove beneficial rather than harmful.

The fraction of global warming skeptics among scientists who study climate change has been disputed, partly because there are various degrees of skepticism. The extreme skeptics are certainly in the minority among mainstream scientists. George Mason University climate researcher Jagadesh Shukla claims to have come across no more than a handful of skeptics among the hundreds of active workers in the field.[8] Whatever the relative numbers on each side, that comparison says little about the validity of their respective positions. Science is not a "majority rules" enterprise.

One skeptic is climatologist Patrick Michaels of the University of Virginia, who prefers to think of himself as a "data-driven moderate." Michaels is the coauthor of the wittily titled book *The Satanic Gases*, in which he argues that given the large rise in federal funding for the subject of climate change (a thousandfold in the last 15 years), and the biases of many climate scientists, dissenting views such as his tend to be suppressed.[9]

After all, argues Michaels, why should scientists support research that attempts to show that global warming may not be a serious problem and therefore need not be the recipient of federal funds? Michaels' argument has some validity, but probably not as much as he thinks. Few self-respecting scientists would want to work on a nonproblem just to stay on a "gravy train" of federal grants. And still fewer scientists would restrain their

145 natural tendency to look for flaws in prevailing theories because of political or other biases. (As a colleague of mine has noted indignantly, "Scientists don't conspire, they compete!"[10])

Michaels imputation of bias can also be turned around and aimed at the skeptics too. It is clear that there are conservative political forces that tend to amplify the voices of the skeptics in order to derail efforts to curb global warming. In addition, through their dissent the skeptics are able to sell books, get private funding, and gain considerable media attention. In fact, there is probably some truth to the arguments on both sides. You can judge for yourself which side has the better case regarding their perception of bias on the other side. Our main interest here will be in assessing the validity of the evidence presented on each side. In particular, how good is the case for not worrying too much about global warming?

How Much Global Warming Has Occurred So Far?

Only in the last few centuries has it been possible to make direct temperature measurements using thermometers. However, various other indirect data, including the isotopic composition of trapped air bubbles in Greenland ice cores, allow scientists to infer what the global temperatures have been going back hundreds of thousands of years. Ancient temperatures can be inferred based on the abundance of different isotopes of oxygen trapped in the bubbles, because the relative rates at which those isotopes evaporated from the ocean surface depends on temperature. (For example, oxygen-18, being heavier than oxygen-16, evaporates less readily, but the difference becomes less significant the warmer the temperature, so air bubbles containing oxygen enriched in the former isotope indicate higher temperatures.)

Currently, the Earth is between ice ages, which occur at roughly 100,000-year intervals. During the last ice age polar temperatures were around 10 degrees Celsius colder than now, and the planet is probably now warmer than any time during

146 the last 1000 years. The warming that has occurred since the last ice age was, of course, not human-induced, but rather the result of natural variations. If our interest is mainly in the human influence on climate, we need to examine shorter time spans corresponding to the period of rapid rise in usage of fossil fuels, i.e., the last century, particularly the most recent decades. The graph in figure 6.2 shows changes in the average global surface temperature since 1860. The data were obtained by averaging many daily temperature readings taken on land and sea around the globe. The zero level of temperature changes is an arbitrary reference level chosen as the average temperature between 1961 and 1990. Based on this graph we can infer that global temperatures have risen about 0.6 ± 0.2°C during the last century, with the rise being most dramatic in two time periods, 1910–1945 and 1976 to the present. (Throughout this chapter we'll cite temperature changes in degrees Celsius only. The corresponding changes in degrees

Departures in Temperature (°C)

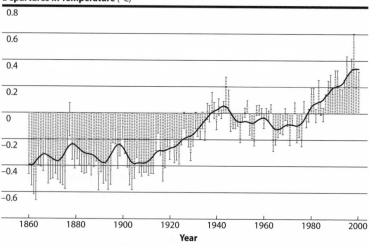

Figure 6.2 Average global surface temperatures since 1860, shown as departures from the 1961–1990 average. From the Intergovernmental Panel on Climate Change.

147 Fahrenheit are 1.8 times as large. Uncertainty estimates cited for temperature rises such as the above ±0.2°C will be given at the 95 percent confidence level.)

An enormous number of individual measurements went into generating the data plotted in figure 6.2. Daily temperature readings made at thousands of weather stations spread throughout the globe, on both land and sea, were averaged, taking into account the greater concentration of stations in some areas than others. Corrections also were applied to these data to reflect different techniques for making the measurements over the years, particularly at sea. Finally, as explained below, an attempt was made to correct the data for the "urban heat island effect," which can give rise to a spurious warming.

The urban heat island effect occurs when a weather station, typically located at an airport, finds itself in an area that over the years has become increasingly urbanized. Land areas covered by concrete and asphalt tend to retain heat much more than rural areas covered by farmland and woodland. Therefore, increased urbanization around a weather station will tend to give rise to higher temperatures over time. These higher temperatures may be accurate for the weather station itself and its immediate surroundings, but they are unrepresentative of the hundreds of square miles of nonurbanized land around it. Some global warming skeptics have suggested that a significant portion of the observed 0.6°C rise in surface temperatures during the last century may be due to the heat island effect.

The data plotted in figure 6.2 have supposedly been corrected for this effect, but let's take a closer look at how such corrections are made by looking at figure 6.3. Generally, what is done is to compare long-term temperature records of several nearby weather stations (A and B), which should pretty well track one another if no significant heat island effect is operating at either one. If station A begins to show systematically higher temperatures than station B beginning at a certain year, then we can infer that the temperature increases were caused by a heat island effect around station A that became significant at that

148 year. As shown in figure 6.3, A and B begin to diverge in the early 1970s. Since 1970 falls well before the end of the record, it is relatively easy to correct for the heat island effect by eliminating readings for station B from the early 1970s on.

Correcting for the heat island effect becomes much more difficult the closer we get to the present. For example, as shown in figure 6.3, station C begins to diverge from A only about 1985, and it is unclear if the degree of divergence at the end of the temperature record is yet large enough to cause station C to be eliminated. Consequently, we can be sure that some bias toward higher temperatures (due to the heat island effect) will remain in the record the closer we get to the present. Unfortunately, the most recent times are when we need to have the most accurate data if we are trying to spot evolving temperature trends.

It is likely that the global temperature record contains some heat island bias, but figuring out how much is difficult. Is it possible that the entire temperature rise since 1975 seen in figure 6.2 is due to heat island bias as some global warming skeptics have claimed? The skeptics are willing to concede that the earlier pre-World War II rise in global temperature was gen-

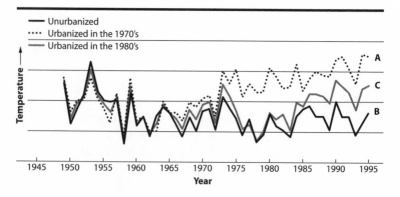

Figure 6.3 Distortion of the temperature record from the urban heat island effect. From Michaels and Balling's *The Satanic Gases*.[9] Printed with permission of the Cato Institute.

149   uine, since, as explained, removing the heat island effect can be
done fairly well except in the most recent years.

One piece of evidence cited by global warming skeptics that
shows that the heat island bias may be producing a spurious
warming during the last several decades is temperature data
recorded by satellites. The satellite record, which goes back to
1979 (see figure 6.4), has its ups and downs, but it shows no
overall trend over the same time period in which measurements of the surface temperature rose by 0.15 ± 0.05°C per
decade. Satellites do not record the temperature of the Earth's
surface. Instead, they measure the temperature of the lower atmosphere, based on the amount of microwave radiation emitted by air molecules in the lowest 8 kilometers of the Earth's
atmosphere.

How do satellites use this radiation to measure air temperature? A body at some given temperature will emit radiation
whose spectrum follows a so-called blackbody shape. The observed shape, including the position of the peak of the spectrum, tells us the temperature of the emitting body, in this case
the Earth's atmosphere. In some respects the satellite measurements, while indirect, are more reliable than the surface data.

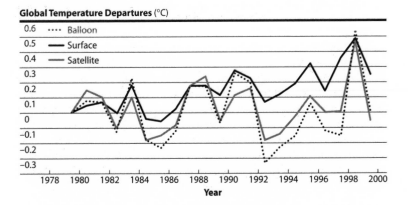

Global Temperature Departures (°C)

Figure 6.4 Satellite, surface, and weather balloon data from Michaels and Balling's
The Satanic Gases.[9] Printed with permission of the Cato Institute.

150 First, satellites "see" a large fraction of the Earth at one time, so there is no question of any sampling error, which could occur for the surface data obtained by averaging temperatures from many isolated weather stations. Specifically, there is no possibility of satellite data showing an urban heat island effect that would yield a spurious temperature rise over time.

The difference between the two time trends for the satellite data, which is consistent with no rise (or at most $+0.05 \pm 0.10°C$ per decade), and the surface data, which show a clear rise ($+0.15 \pm 0.05°C$ per decade), is statistically significant, but only marginally. Global warming skeptics note that the independent record of weather balloon data (see figure 6.4), agrees with the satellite data very well, but not with the surface data. Thus, we see that two out of three independent temperature records made since 1979 show virtually no warming over the last few decades. This observation supports the notion that the surface data may be spurious due to the heat island bias.

Another likely possibility is that the surface temperature data and the satellite data are both genuine, but they are merely recording different temperatures (one on the surface and the other in the atmosphere). In the words of one recent climate report, "It is physically plausible that over a short time period (e.g., 20 years) there may be differences in [the two] temperature trends."[11] One recent study on large volcanic eruptions, for example, suggests that the dust and gases they spew out into the atmosphere can explain why the records of surface and atmospheric temperatures have diverged in the last two decades.[12] It is also possible that the surface temperatures lag those in the atmosphere due to the large heat capacity of the oceans. Whatever the explanation, however, the idea that the surface and atmospheric temperature records are truly going in different directions becomes increasingly implausible the longer the two temperature records diverge. It is simply not possible to warm the surface of the planet without the lower atmosphere also warming up after several decades.

What of the skeptic's claim that much of the recent surface warming is due to the urban heat island effect? The record of

151 nearly equal temperature rises in the two hemispheres of the planet provides strong evidence against this claim. Since 1979, the rise in global surface temperatures appears in both the Northern and Southern Hemispheres: +0.17 and +0.13°C per decade, respectively. The Southern Hemisphere is 90 percent ocean and significantly less populated than the Northern Hemisphere, so it should show much less heat island bias. Since the temperature rises are fairly similar in the two hemispheres, we can infer that the heat island effect is not very significant.

A second piece of evidence against the skeptics is the balloon data, which go much further back than the satellite data—all the way to 1958. Over this interval the balloon data show a rise in atmospheric temperature of +0.09°C per decade, which is very close to the surface data rise of +0.10°C per decade. That close agreement suggests that the surface data are *not* in error, and that over a more extended period of time than the last two decades the surface, satellite, and balloon data sets should agree. In other words, it would appear that the surface really has warmed since 1979, but the atmosphere has not yet caught up due to its temporary cooling because of large ocean heat capacity, volcanoes, or other factors.

Many other indicators also seem to point to a real global warming in recent decades, including the retreat of glaciers, a continuation in the rise of sea level (about 1.5 centimeters per decade), and an observed rise in the heat content of the world's oceans since the 1950s.[13] There is one final piece of evidence that indicates that the Earth really has warmed over the last few decades. The balloon data show that since 1958 the temperatures in the lower stratosphere (the part of the Earth's atmosphere above 10 kilometers) have been dropping over time by about 0.15°C per decade.

It may seem strange that a cooling stratosphere constitutes evidence for a warming surface, but that is exactly what would be expected if the cause of the warming is an increase in greenhouse gases. Overall, the balance of the evidence is in favor of a real global warming taking place over the last few decades. If

152 that conclusion is correct, the surface, balloon, and satellite temperature data should agree better as time goes on.

How Sure Is It That Rising Global Temperatures Are Anthropogenic?

Not all of the greenhouse effect is human-caused. In fact, a sizable natural greenhouse effect is important in making the Earth a habitable place. Without any greenhouse gases in the atmosphere temperatures on the surface would be approximately 33°C colder (well below freezing), and life might not be able to flourish or even survive. The basic idea of the greenhouse effect is similar to what happens in a greenhouse or a car with locked windows on a sunny day. Incoming solar radiation in the form of visible light easily enters the car's windows and warms the interior. The warmed interior emits infrared (heat) radiation, but it cannot escape easily because it is blocked by the glass windows, which are fairly opaque to infrared radiation generated by the warmed interior. Eventually, the interior gets so hot that even though most of the infrared radiation cannot get through the windows, enough does get out so as to balance the incoming energy in the form of visible light.

In the case of the atmosphere, the greenhouse gases play the role of the car windows: they allow light from the sun to come in, but prevent much of the infrared radiation from the warmed Earth to escape into space. In this respect the greenhouse gases act like an insulating blanket on the Earth. The thicker the blanket of gases, the more the Earth's surface is warmed. The blanket analogy is faithful in another respect. Like a real blanket, the thicker you make it, the *cooler* it will be on the outside surface of the blanket, since the heat from the warmed Earth doesn't escape as easily. This observation explains the earlier strange comment about how rising global surface temperatures due to the greenhouse effect are associated with cooling of the stratosphere.

One other piece of evidence in favor of greenhouse gases

153 being a significant cause of planetary warming is the close connection between atmospheric levels of greenhouse gases and the global temperature at various prehistoric times. Note, in particular, the times at which the most prominent peaks and valleys occur in the separate records of temperature, CO_2 and methane in figure 6.5, which strongly suggests a connection between these two greenhouse gases and temperature.

Unfortunately, parallel time trends between A and B cannot establish that A is the cause of B. While it is certain that some amount of greenhouse effect has been caused by changes in atmospheric CO_2 and methane levels, it is possible that the reverse direction of causation is more significant. (Reverse causation implies that changing temperatures lead to changing atmospheric levels of greenhouse gases.) That scenario is possible because the oceans represent a tremendous repository of dissolved gases, and as the oceans heat up they release greater quantities of these gases into the atmosphere.

In fact, one analysis of temperature changes at the end of the last three ice ages claimed that temperature rises *preceded* CO_2 rises by 600 ± 400 years.[14] But, the authors of that analysis made no claim that higher CO_2 levels were the result of higher temperatures. Ultimately, the observed close connection between atmospheric greenhouse gas levels and global temperatures for the ancient Earth offers little evidence on the direction of causation, unless the uncertainty (± 400 years) should shrink considerably. More importantly, those data offer no evidence against the greenhouse effect—an effect that is based on well-established physical principles, which even the skeptics accept.

There are many reasons the Earth might heat up or cool down aside from the greenhouse effect. In general, anything that disturbs the balance between the incoming solar radiation and the infrared radiation that the warmed Earth sends back out to space will have such an effect. Volcanoes, for example, as already noted, represent one such disturbance, since when they throw a huge volume of dust and aerosols up into the air they have the effect of blocking sunlight, and therefore cooling the Earth. (Just think how much cooler it is in the shade than in the

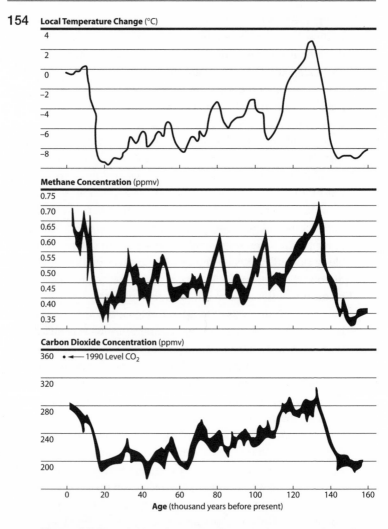

Figure 6.5 Record of ancient temperatures and concentrations of two greenhouse gases. From Houghton's *Global Warming: The Complete Briefing*.[54] Reprinted with the permission of Cambridge University Press.

155 direct sunshine on a hot summer day.) Another cause of a changed energy balance would be if the sun were to increase or decrease its output slightly over time. In fact, it is believed that billions of years ago the sun was fainter than at present, and a natural greenhouse effect was even more important then for the evolution of life than now.

Could the Sun's Changing Output Be the Cause of Global Climate Change?

Some researchers have suggested that the global temperature variations since 1860 are entirely due to changes in solar activity. Even though direct measurements of the sun's output at visible wavelengths don't seem to show enough of a variation to be responsible for a significant portion of the Earth's climate change, some researchers argue that some amplification mechanism could be at work.[15] (The rationale for that suggestion is that global temperature decreases are surprisingly large during periods of decreased solar activity.)

Moreover, even if the sun's visible output doesn't change appreciably over time, one solar feature does change dramatically, the number of sunspots. Sunspots can be seen and counted easily using a modest telescope, and records of their number have been kept for centuries. Sunspots, whose number goes through a well-known eleven-year cycle, have been considered responsible for all manner of Earthly events. At the peak of the sunspot cycle, Earth experiences geomagnetic storms and communications disruptions, so sunspots do have real effects on our planet. Some of the phenomena claimed to be affected by sunspots are fanciful (the stock market or the length of skirts), but others, such as the Earth's climate, may not be.

As figure 6.6 shows, the cycle in the annual number of sunspots is quite irregular, in terms of its shape, maximum height, and period (cycle length). In a 1991 article, K. Lassen and E. Friis-Christensen, two scientists with the Danish Climate Center, observed a remarkable degree of correlation be-

156 tween global temperatures (solid curve in figure 6.7) and sunspot cycle period (dotted curve).[16] In this case, if one of these phenomena is really causing the other, there is no way for causation to run in the "reverse" direction, i.e., for changing Earthly temperatures to cause changes in the length of the sunspot cycle. Therefore, global warming skeptics have pointed to this connection between the length of the sunspot cycle and the average surface temperature as evidence that changing temperatures could be caused by the sun rather than the greenhouse effect.[17]

There are three serious flaws in the skeptic's argument that the cause of global warming is due to variations in sunspot cycles. The first flaw is that the degree of match between the two time trends is actually less impressive than it looks in figure 6.7. Bear in mind that the right vertical scale (for temperature) and the left vertical scale (for length of cycle) are separately chosen to make the two time trends have the best agreement. *The proper choice in the two scales virtually guarantees that the two time trends can be made to agree somewhere near the beginning and end of*

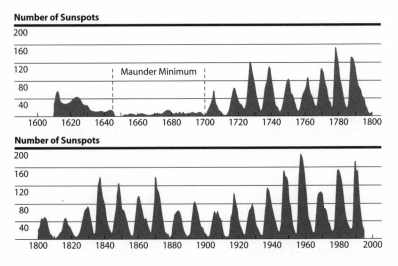

Figure 6.6 Number of sunspots versus time, courtesy of the National Geophysical Data Center, Boulder, CO.

157 *the time interval plotted.* A second flaw is that there are other sunspot variables, such as the maximum height of each cycle, the width of each peak, or the total number of sunspots per cycle, that might have been used. Any of these variables might have shown a correlation, but were apparently not used because the cycle length evidently showed a better fit to the temperature data.

Third, because the cycles are irregular, the definition of the sunspot cycle length (i.e., the period or length) is not unique and involves various possible types of averages over successive cycles. The method of averaging that gives the best fit shown in figure 6.7, for example, involves defining the length of any given cycle as the weighted average of five cycles centered on the cycle of interest. In other words, the cycle length plotted for the year 1990 depends in part on the cycle length two cycles (22 years!) on either side of it. Without such a dubious averaging— in effect a smoothing over five consecutive cycles—the length of each cycle would be much too irregular to give any kind of decent fit to the temperature data.[18] Given all the choices to be made in a particular sunspot cycle variable and its exact defini-

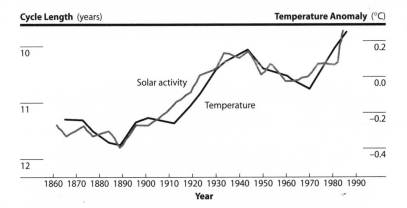

Figure 6.7 Correlation between length of sunspot cycle (light curve) and average global temperature (dark curve). Reprinted with permission from Friis-Christensen and Lassen.[16] Copyright © 1991 American Association for the Advancement of Science.

158 tion, it is not surprising that one of the choices should show a good correlation with global temperature, when plotted in just the right way. (Recall the pitfall of making "informed choices" discussed in chapters 2 and 5.)

In a 1999 paper Thejil and Lassen updated the earlier 1991 study on the connection between sunspot cycles and global temperatures.[19] The more recent data show that the fit for the years since 1991 is not nearly as good, with actual global temperatures during the last decade being significantly higher than what would be predicted based on the sunspot cycle length. Thejil and Lassen's interesting conclusion is that the higher than expected temperatures observed in the 1990s are a possible indication of human-caused activity, i.e., the effect of greenhouse warming.

Another more plausible interpretation of the 1999 result is that the earlier-reported correlation between sunspot cycles and global temperatures was just a statistical fluke, since the newly added data no longer fit the earlier pattern. Whichever explanation is true, however, the global warming skeptics lose points. In other words, anthropogenic global warming becomes more plausible if either the departure in the recent data is due to greenhouse gases, as Thejil and Lassen suggest, or the earlier correlation between sunspots and global temperatures was just a fluke.

The preceding analysis suggests that the link between the length of the sunspot cycle and global temperature during the last century is tenuous and probably incorrect. On the other hand, notwithstanding this dubious sunspot study, there *does* appear to be a real link between solar output variations and changes in the Earth's climate. One recent study looked at this possibility by studying the relationship between carbon-14 and oxygen-18 isotopes over a 3000-year period starting about 10,000 years ago, as recorded in a stalagmite found in a cave in Oman.[20] The idea goes like this. ^{14}C is a radioactive carbon isotope that is created by cosmic rays, some of which originate in the sun. Therefore, the ^{14}C level of radioactivity is indicative of solar activity. Furthermore, the ^{18}O level is indicative of ocean

159 temperature, because, as already noted, the heavy oxygen isotope evaporates more easily, the warmer the ocean. The observed close connection between the time trends for these two
isotopic concentrations provides evidence of a link between solar activity and global temperature.

Still, there is a real question of whether a possible link between solar activity and global climate passes the "so what?"
test. Remember that our main interest in the preceding section
is in evaluating the case for a human influence on climate. The
question of whether human activity influences climate cannot
be resolved by proving that solar activity also influences climate. To answer that question we need to turn to computer
models of the greenhouse effect.

Global Climate Models Based on the Greenhouse Effect

The usual way of testing whether rising temperatures are
caused by atmospheric greenhouse gases is to run a computer
model and see how well the model results fit the observed temperature data. These general circulation models (GCMs) essentially are the same kind of models used to make weather fore-
casts. GCMs divide the Earth's surface into a two-dimensional
grid of cells along lines of latitude and longitude. Within each
two-dimensional cell the atmosphere above that cell is then divided into a number of vertical layers, effectively dividing the
entire atmosphere into a large number of boxes.

For each box all the relevant data (temperature, pressure, hu-
midity, wind speed and direction, etc.) must be supplied at one
particular starting time. The model is then allowed to run one
step at a time to see how the conditions in each box change over
time, as air masses in each box interact with those in adjoining
boxes and with the Earth's surface. GCMs are complex and require long times to run, even on the most powerful computers,
primarily because of the large number of boxes and the large
number of time steps needed to advance the model results
years into the future.

160   You might wonder about the plausibility of such computer models being able to predict the climate years into the future, given the well-known difficulties of predicting the weather just days or weeks ahead. But keep in mind there is a big difference between weather and climate. (It has been said that climate is what you expect, but weather is what you get.) It's much easier to predict that a given region is likely to experience a hotter or wetter summer (climate) than it is to predict whether it will rain or snow in Chicago on a given day two weeks from now (weather).

It's also much easier to predict whether the Earth as a whole is likely to experience higher temperatures in the future than whether any given location will. Finally, we are, for the moment, considering only the ability of the GCMs to "predict" the past, which is much easier to evaluate than their ability to predict the future! In other words, if models are started out with conditions as they were back in 1860, how well do the year-by-year average global temperatures they generate agree with the already observed record? In "predicting" the past, it is always possible to fine tune the models to give better and better agreement, since we know what the answers are. Keep in mind that finding a good fit to the observed temperature record as a result of such fine tuning could give us an unwarranted degree of confidence in the correctness of the models. (Physicists have a saying that with enough free parameters, you can fit an elephant i.e., make any weird set of data fit a theory.)

Also bear in mind that various researchers each have their own models, which give a range of results, depending on the specific assumptions that are made. The Intergovernmental Panel on Climate Change (IPCC), comprised of scientists from 99 countries, has compiled and synthesized the results from various models and released a series of reports that are published under the auspices of the United Nations. To date, three main reports have been released by the IPCC, the third one published in 2000 with 515(!) contributing authors. A figure from the latest IPCC report, reproduced as figure 6.8, shows how well and under what assumptions the models reproduce the global temperature record since 1860.

161   According to the two graphs in figure 6.8, with only natural fluctuations (top graph), the models fail to reproduce the rise in temperatures seen over the last few decades. Natural variations include both the effects of volcanoes and slightly varying solar output, but not any solar variations associated with the speculative idea discussed earlier that sunspot cycles affect climate. The best fit to the observed temperature record (bottom graph) is found when the models include a combination of natural variations and anthropogenic (human-caused) effects. Given the significant spread in the predictions of different models, in-

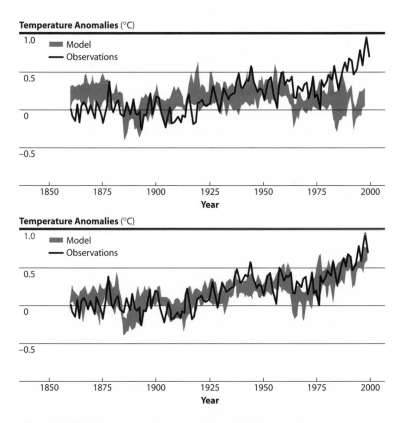

Figure 6.8 Fits to the temperature record since 1860: top graph includes only natural variations; bottom graph includes both natural variations (such as volcanoes) and anthropogenic sources. From the Intergovernmental Panel on Climate Change.

162 dicated by the width of the gray band in the figure, the models are quite consistent with what has been observed, but the previous comment regarding the pitfalls of "fine tuning" should be kept in mind. The absence of equally good fits to the temperature records in the separate hemispheres offers further reason for caution.

What about Sulfate Aerosols?

In addition to greenhouse gases, the human-caused effects also include the effects of "sulfate aerosols." Aerosols are microscopic airborne particles, whose main anthropogenic source is the burning of fossil fuels. Without the inclusion of sulfate aerosols the models tend to predict more warming than is actually observed, because these aerosols have the effect of blocking sunlight and canceling out part of the greenhouse warming. Climatologist Patrick Michaels regards the inclusion of sulfate aerosols in the GCMs as sort of a "fudge factor" that has been added to fix up otherwise deficient models.[21]

Scientists on both sides of the debate agree that the effects of the aerosols are extremely uncertain, especially their indirect effects in forming clouds, which could have an impact on moderating greenhouse warming that ranges anywhere from zero to very major. According to the skeptics, models that include such aerosols are virtually guaranteed to fit the data, depending on what assumptions are made regarding their cloud-forming properties. Patrick Michaels also notes that one important aspect of the greenhouse plus sulfate model seems to be at variance with observations, namely the size of the expected warming in the two hemispheres. Given that most of the fossil fuel usage occurs in the Northern Hemisphere (NH), Michaels suggests that it should show a smaller temperature rise in recent decades than the Southern Hemisphere (SH), which is just the reverse of what has been found. In fact, during the last half-century the observed surface warming has been "almost exclusively confined to the dry cold anticyclones of Siberia and North America."[22]

163   Michaels' observation that the NH should show the smaller temperature rise fails to take into account several offsetting factors. First, if there really is some heat island bias acting to give spuriously higher surface temperatures in the NH, that bias might easily explain the difference in the two hemispheres. For example, if a third of the decadal temperature rise in the NH were due to the heat island bias, then the actual temperature rise in the SH would exceed that in the NH. A second offsetting factor is that Michaels considers merely the levels of sulfate aerosols in the two hemispheres, and not their rates of change. In recent decades tighter restrictions on emissions from coal-burning plants have had a much greater impact on aerosols in the NH, which had much higher levels to begin with. Sulfate atmospheric levels have dropped more than 25 percent since 1970, based on measured deposits in Greenland ice cores.[23] Therefore, we might expect that the temperature rise in the NH in the last three decades actually should be greater than in the SH, because reduced sulfate levels don't cancel out as much of the greenhouse effect there as they did in 1970.

Aside from hemispheric differences in global warming, are there other indicators that might suggest whether or not the observed warming in recent decades is due to the greenhouse effect? One indicator noted earlier was the *drop* in stratospheric temperatures—remember that as you add on more blankets on a cold winter night the outer (top) blanket gets colder, i.e., closer to room temperature, while you get warmer under the blankets. Another indicator that the warming is due to the greenhouse effect is its temporal pattern, which has mainly been during cold winter months, rather than warmer months. (For example, during the last 50 years, over two-thirds of the warming that has taken place in the Northern Hemisphere has been in the winter months.[24] Moreover, the greenhouse effect predicts that day–night minimum temperatures should increase faster than maximum temperatures. In fact, this narrowing of the daily temperature range has occurred in recent decades, with the minimum temperatures increasing twice as fast as the maximum.[25]

164   When all the evidence on both sides of the issue is taken together, it appears that those who worry about global warming
have a better case than the skeptics regarding the cause of the
warming that has occurred in the past few decades. In their second assessment report (1995), the IPCC noted that "the balance
of the evidence suggests a discernible human influence on
global climate." To a physicist, or at least this physicist, a "balance of the evidence" type of statement sounds like a fairly
modest degree of certainty, of perhaps one standard deviation
confidence. The third (2000) report makes the somewhat
stronger claim that "there is new and stronger evidence that
most of the warming observed over the last 50 years is attribut-
able to human activities." Both the 1995 and 2000 IPCC claims
appear to be warranted, although each is capable of being interpreted as conveying a greater degree of certainty than the
IPCC may have intended.

What Is Likely to Be the Extent of Global Warming by 2100?

Global warming skeptics usually don't claim that there will be
zero warming in the 22nd century, only that it is likely to be
very small. For example, S. Fred Singer, author of *Hot Talk, Cold
Science: Global Warming's Unfinished Debate,* probably can be
classified as being an extreme skeptic.[26] In his book Singer suggests that the average global temperature is likely to rise only
about 0.5°C by the year 2100. Singer, who believes that the
record of surface temperatures over recent decades is unreli-
able due to heat island bias, obtains his estimate from the satellite record, which shows very little warming (0.05 ± 0.10°C per
decade since the record began in 1979). Singer assumes that this
increase of 0.05°C per decade could be extrapolated for the
coming ten decades, and obtains his estimate of a temperature
increase of perhaps 0.5°C for the year 2100. Although Singer
doesn't quote any uncertainty on this projected temperature increase by 2100, he would presumably say that the likely in-

165 crease would be in the range 0.5 ± 1.0°C, i.e., at most 1.5°C, since his estimate for 2100 is based on the satellite trend data.

Patrick Michaels also believes that the record of surface temperatures is less reliable than that of satellites, but he is willing to use the surface data to give his projection for the year 2100. Based on the decadal rise (0.15 ± 0.05°C) seen in the surface data, Michaels projects that trend ahead ten more decades to obtain a rise of ten times 0.15 or 1.5°C by the year 2100, which is still a fairly modest increase, compared to values suggested by the IPCC.

According to Michaels, as long as greenhouse gas concentrations are assumed to be increasing at the same fixed percentage per year, the temperature rise will continue to be linear. The reason is that each added increment of greenhouse gas to the atmosphere has less of an effect than previous ones—in the same way that on a very cold night you'd need to pile on two more blankets to have the same effect as the first one, and four more after that.

Is Michaels right that future temperatures are bound to rise at a constant linear rate, assuming a fixed annual percentage rise in atmospheric greenhouse gases? Not necessarily. Many nonlinear effects exist in climate models dealing with cloud formation, and some unknown mechanisms might also introduce nonlinearities. For example, as much as half of the CO_2 emitted each year does not go into the atmosphere, but instead is sequestered in various carbon "sinks," including the world's oceans. Up to half of the removed carbon is going into some sink that has not yet been identified. It is conceivable that at some point that unknown carbon sink could become saturated and unable to absorb more carbon, causing global temperatures to rise significantly faster than at present.

It also seems implausible that scientists would bother with all their fancy climate models if an elementary school student with a ruler could simply do a linear projection from past trends to find the likely average global temperature at any future date. Obviously, making future projections is not quite as

166 simple as some global warming skeptics suggest. The contrary observation of skeptics that climate researchers need to justify the continued flow of research dollars and to use methods that may be more complex than necessary seems less valid, but it may be not entirely without merit.

Figure 6.9 shows IPCC global temperature projections for the coming century. Unlike the skeptics who suggest a specific value (with no uncertainty) for the likely warming during the coming century, the IPCC shows a wide range of possibilities, and these rising temperatures obviously become more uncertain the further into the future they project. Taking the width of the uncertainty band, the IPCC projects a temperature rise by the year 2100 of anything from 1.4 to 5.8°C higher than present. This huge range would correspond to between 2.5 and 10.4°F. However, remarkably, no indication is given in the IPCC report as to what this uncertainty range represents. Is the likelihood of exceeding the quoted range one in ten, a thousand, or a billion?

The huge range in the IPCC-predicted rise by 2100 has two sources of uncertainty: future emissions of greenhouse gases,

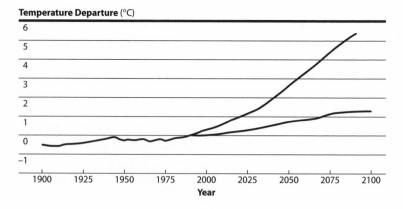

Figure 6.9 Global temperature changes through 1998 and the projected global temperature rise through 2100. The diverging curves after 1998 show the upper and lower bounds on the uncertainty range due to both differing models and emissions scenarios. From the Intergovernmental Panel on Climate Change.

167 and the correctness of the model to make the projections. Let's start with the first source of uncertainty. It is obvious that without knowing the amount of future greenhouse gas emissions, no one can predict how the level of atmospheric greenhouse gases will change in the coming century. It is probably naïve to assume that these emissions will continue to increase at the same fixed percentage that has prevailed in recent decades.

To deal with the uncertainty in emission levels the IPCC considers a wide range of different scenarios, which depend on possible future economic growth and fossil fuel usage in various parts of the world. These scenarios give results lying between the curves in figure 6.9. Nearly all of them give much higher estimates of the temperature rise than the skeptics suggest, because the scenarios mostly involve higher growth rates in CO_2 emissions than have prevailed in recent decades.

The main reason for the rapid growth in CO_2 levels in the most "pessimistic" scenario (the uppermost curve in figure 6.9) is that this case assumes that the developing world will rapidly catch up to countries in the West economically and will continue to cut down its forests. It is also assumed that these newly rich countries will emulate the West in terms of its very heavy dependence on fossil fuels and its lack of concern over energy efficiency. Clearly, once the populations of all the nations of Asia and Africa are driving SUVs, CO_2 emissions will skyrocket. (In the "gloomiest" of the IPCC scenarios, atmospheric levels of CO_2 in 2100 are assumed to be 340 percent of their present value. That figure would represent an annual rate of increase of 1.2 percent, which is quadruple that seen in recent decades.)

Quotes were used in describing the highest CO_2 emissions scenario as "pessimistic" because it seems unlikely that people in the developing nations would regard the underlying assumption of rapid economic convergence with the West as anything other than a highly positive development, regardless of the environmental impact. However, the realism of that rapid economic convergence scenario is open to question, even though it may be politically incorrect to say so.

168   Individual scientists who do the actual climate modeling may be well aware that the rapid economic convergence emissions scenario is extremely unlikely, and they may be unafraid to say so. But, the IPCC, an international committee representing 99 nations is unlikely to dismiss the possibility that countries in the developing world will soon catch up to the West. The most pessimistic scenario is unlikely not only on geopolitical and historical grounds, but on economic grounds as well. If a rapid economic Third World catch-up should occur, it is unlikely to rely heavily on fossil fuel use, because market forces would probably lead to the abrupt end of cheap petroleum. For example, according to geophysicist Kenneth Deffeyes, the annual world production of oil will peak in a few years, with major disruptions in supply and rising costs likely to follow.[27] Such a scenario of declining oil production would be entirely incompatible with rapidly rising Third World living standards that depend on fossil fuels.

On the other hand, categorizing the most extreme fossil fuel scenario as being extremely unlikely should not imply that it is unworthy of consideration. There is still utility in using extremely pessimistic, albeit unlikely, scenarios to look at how the global climate might change, because they help us get a sense of what is likely to be the worst that might occur. Still, we do need to keep in mind that all scenarios are not equally likely when we consider the uncertainty in future global temperatures dependent on the choice of CO_2 emissions scenario.

The second source of uncertainty that gives rise to the enormous range of IPCC temperature projections for the year 2100 is that different computer models give different results for the same assumed increase in the level of atmospheric greenhouse gases, i.e., different "climate sensitivities." One of the biggest sources of variation between the different models is how clouds are treated, since different assumptions about clouds can lead to either a surface warming or a surface cooling. For example, low-altitude cumulous clouds cool the surface by blocking sunlight, while high-altitude, wispy, cirrus clouds warm the surface because they let sunlight through but tend to keep infrared radiation from escaping.

169   Clouds are especially difficult to deal with because they form on a spatial scale that is much smaller than the size of the boxes used in the models. This uncertainty concerning clouds is a special case of the two types of global warming feedback: the positive types that amplify any warming, and the negative ones that reduce it. Atmospheric water vapor, for example, when it *doesn't* form clouds is a source of positive feedback, because when the planet warms, more water evaporates, and water vapor, being the dominant greenhouse gas, causes still more warming. Although the contributions from various feedbacks depend significantly on the particular model, on average their net result is to amplify the direct greenhouse warming by 250 percent. *A major reason for the differences in model predictions is that some models give greater emphasis to positive feedbacks, while others give greater emphasis to the negative ones.*

All these uncertainties in the models limit our ability to make future projections, even for a particular emissions scenario. On the other hand, global warming skeptics argue that nature has already answered the question regarding which models are correct, based on which ones fit the linear rise in temperature seen during the last few decades. Therefore, the skeptics say that including the full range of models to make future projections is very questionable.

Conversely, other scientists, such as James Hansen, director of the Goddard Institute for Space Studies, argue that we should retain a wide range of models in making future projections. While it is true that seen in the surface data a greenhouse warming signal has been tentatively established based on a linear rise in temperatures seen in the last decade, it is premature to use the slope of that trend to rule out models that don't fit the data. We cannot exclude the possibility that the linearity of the recent temperature rise is just a statistical artifact. Therefore, a narrowing of the range of models used to make projections might be appropriate only after we have another decade's worth of data, according to Hansen.[28]

Both sides of the debate have their valid points here. The skeptics may be wrong to cite only a single value (with no uncertainty) for the expected warming by 2100, and also to simply

170 assume that a linear trend over the last 20 years can be extrapolated forward for the next century. Conversely, the IPCC can be faulted for giving too great a range of possible values, i.e., too large an uncertainty range, without making any judgments on the likelihood of the extreme temperature predictions.

In fact, Stephen Schneider, professor of biological sciences at Stanford University and lead author of the IPCC paper on uncertainties has been critical of the IPCC report on precisely this last point. Schneider argued unsuccessfully for including in the IPCC report expert estimates of the likelihood of various projected rises in temperature, being fully aware of the contentious nature of the debate that this would entail. Without such expert estimates, Schneider surmised that people would conclude falsely that all values of the projected temperature rise suggested by the IPCC were equally likely.[29]

Some of the reluctance of the IPCC to confront this difficult issue of assigning probabilities to various temperature rises may be the desire of scientists to avoid a debate that could prove fruitless. However, one need not be a global warming skeptic to wonder whether there may have also been political reasons for the wide temperature range cited by the IPCC, on the part of some scientists. Scientists would be less than human if they did not occasionally let their beliefs color their scientific judgments. For example, a large group of scientists compiling an overall assessment based on their various models may seek to achieve group cohesiveness by not excluding some models as being less realistic than others. In addition, an international committee representing scientists from 99 countries may not wish to offend the sensibilities of colleagues from developing nations by implying in print that some emissions scenarios (associated with very rapid Third World economic growth) are unrealistic.

Finally, those scientists who believe that global warming is an urgent problem that needs much greater public attention may believe that the way to get that attention is to include a very high upper bound to the temperature increase. In that way, the media will likely focus on that figure, i.e., "the temper-

171 ature increase by 2100 could be *as high as* 5.8°C or 10.4°F," so that people will be more likely to take the problem seriously. In addition, with a very high upper bound to the temperature increase, the implications for the environment become more serious, and hence more worthy of study.

A Case of Bias or Character Assasination?

Stephen Schneider, a prominent figure in the global warming debate, knows well how charges of bias can arise and how difficult they are to refute once they do. In a 1989 issue of *Discover* magazine, Schneider is quoted to the effect that "we have to offer up scary scenarios. Make simplified dramatic statements, and make little mention of the doubts we have. Each of us has to decide what the right balance is between being effective and being honest."[30]

This quote has been repeated hundreds of times in articles by conservative authors seeking to discredit Schneider and other scientists working on global warming. The context of the quote was a discussion of the complexities of issues like global warming, and the desire of those in the media to avoid shades of gray and portray this complex issue as though there were only two extreme sides: "good for you" versus "end of the world." Given this media mindset, a scientist can feel a lot of pressure to suppress all reference to uncertainties and present the worst-case scenario that will counter what an antagonist is saying and fit within the allowed 20-second sound bite. The preceding message is what Schneider was trying to convey in his widely misinterpreted *Discover* quote, as he spells out in greater detail in a book chapter on "Mediarology."[31]

Commentators using the quote to demonstrate the bias of scientists always neglect to include the following sentence of Schneider's interview in which he notes that "I hope that means doing both." In other words, it is important that scientists are being both effective and honest, i.e., "leaving the scary stories in because they could happen, but also with honest mention of their likelihood—and the "good for you" extreme as well."[32]

172   So, which side of the debate is likely to be right regarding the warming in the coming century? As noted, the IPCC makes no attempt to assign probabilities of finding different values within their very wide range of temperatures projected for 2100, i.e., 1.4 to 5.8°C, since that would force them to choose which of their CO_2 emission scenarios and models are more realistic. One study published in July 2001 by Wigley and Raper does attempt to assign probabilities, and finds that values near both ends of the IPCC range are very low and that the most likely rise by the year 2100 is 3.0°C.[33] However, Wigley and Raper's study still doesn't assign probabilities to the various emission scenarios, and considers them to be all equally likely.

Let us here, however, make the judgment that the "pessimistic" emissions scenario (340 percent increase by 2100), based on an assumed rapid economic convergence between developing and developed nations, is unrealistic, based on historical, geopolitical, and economic reasons. In that case, the temperature rise by 2100 would more likely be in the lower half of the IPCC range, not far from where some of the skeptics place it. In any case, it is clear that given all the uncertainties, predicting the average global temperature in the year 2100 hinges perhaps as much on the political judgments regarding future Third World economic development as it does on the basic climate science.

Will Rising Global Temperatures Be Harmful or Beneficial?

Global warming skeptics who concede that some warming is likely to take place during this century argue that it is likely to be beneficial rather than harmful. Clearly, the validity of that claim depends on the extent of the warming and how far into the future we are making the projection. It also depends on whether we are considering the effects on humans versus natural ecosystems, and which particular group of humans or ecosystems we wish to focus on. For example, global warming would clearly benefit the residents of a frozen Siberia, but it

173 could be more than a minor inconvenience for some other regions of the globe. Actually, even for the frozen parts of the planet, warming would not be an unmixed blessing, since thawing would lead to such problems as the buckling of paved roads, the sinking of structures built on frozen tundra, and the disruption of wildlife habitats.

Before considering whether global warming might be beneficial on balance, we should first stress that unless you happen to believe that the present global average temperature is "just right," the world might indeed benefit from a modest degree of either warming or cooling. That being said, for the world as a whole, it is likely that slightly warmer temperatures would be better than slightly cooler ones. As already noted, different parts of the world would experience different degrees of benefit or harm if temperatures were to rise, but the notion that overall a warmer world would be desirable is certainly not a ridiculous notion. While it seems ludicrous for humanity to try to achieve artificially the modest degree of warming that might bring the world to some optimum temperature, it seems equally ludicrous to take costly measures to avoid a modest degree of warming, if, in fact, we are currently below that optimum temperature.

Some of the most important ways that rising global temperature could affect the world include impacts on human mortality, agriculture, fresh water availability, frequency of extreme events (storms, floods, and droughts), sea level rise, and natural ecosystems. These impacts include those that are quantifiable in monetary terms as well as some that are not quantifiable (human mortality and ecosystems). In the limited space we have here we will focus on the preceding six categories of impacts; various economic studies have looked at a variety of other areas, including outdoor recreation, timber resources, and energy costs.

Human Mortality. Human mortality rates are directly linked to temperature. While extreme heat and extreme cold are both causes of excess mortality, extreme cold is the more significant

174 cause, killing twice as many people as extreme heat.[34] (For example, the U.S. winter mortality for all causes is currently higher than that in summer by about 16 percent.) Similarly, a 2001 study of mortality in Britain revealed a linear increase as temperatures fell from 15°C to near 0°C, while mortality rates at temperatures above 15°C showed no clear trend. Indeed, low temperatures were shown to have a significant effect on mortality—both immediate (1 day after) and long-term (up to 24 days after) temperature drops. The authors of this study suggest that cold temperatures are deadly because "cold causes mortality mainly from arterial thrombosis and respiratory disease."[35]

Part of the reason that rising temperatures due to the greenhouse effect could reduce mortality is explained by the expected seasonal variation of the temperature rise. Recall that winter warming is expected to significantly exceed that in the summer. In other words, the benefit due to a reduction of extremely frigid weather should outweigh the detriment due to the smaller increase in very hot days. It is also significant that greenhouse warming is likely to be greater in very cold regions of the globe than in hot ones, which should give a further benefit for the world as a whole.

There are, of course, some ways that rising temperatures are likely to increase human mortality, for instance, by causing a possible increase in the spread of diseases such as malaria, due to slightly higher precipitation levels. Malaria now infects 300 to 500 million Africans and kills a million every year.[36] It also seems likely that many of the populous developing nations near the equator would experience higher mortality from rising temperatures. What about the world as a whole—would there be a net rise or decline in human mortality? The IPCC report acknowledges that "limited evidence" indicates that the reduction in cold weather deaths "in some countries" is likely to exceed that of hot weather deaths, but that published studies have so far been done using data from developed countries, thereby "precluding generalizations."[37] This assessment may be technically correct, but it also has the flavor of one that is

175 carefully crafted so as not to give offense to researchers from developing nations, and it could be too pessimistic when one is interested in a net worldwide impact.

Sea Level Rise. One feature of global warming that is of considerable public concern is the rise in sea level, especially if you happen to live in a coastal area. Aside from flooding and storm damage, sea level rise also contributes to pollution of freshwater aquifers, and the loss of coastal property. In fact, however, rising sea levels are not new to the last few decades of human-caused global warming. Sea level has been rising for many centuries, going back to the last ice age, when sea level was an amazing 100 meters below the present value. This figure may be contrasted with what the IPCC projects for the year 2100, i.e., a further rise of between 0.09 and 0.88 meters. This extremely wide range of projected rises in sea level is due to uncertainties in both future emissions scenarios and the models themselves.

You might wonder why the range of possible rises in sea level projected for the coming century is so much larger than the range in projected temperature rises. One reason is that considerable uncertainty exists over the current rate of sea level rise, and the IPCC even declined to give a best estimate for the rise that has occurred during the past century. The two reasons why sea level would be expected to rise with rising temperatures are the melting of glaciers and the thermal expansion of the oceans. There is also an offsetting factor that could cancel out part of the sea level rise. A warmer world is likely to be one with more precipitation. If some of that extra precipitation is deposited as snow in polar regions the moisture would be locked up there and removed from the oceans.

Skeptic Fred Singer believes that this offsetting factor could even be enough to possibly lead to a *drop* in sea level in a slightly warmer world.[38] The only evidence Singer offers in support of this controversial claim, however, is an observed inverse correlation between global average temperatures and sea level. But Singer chooses to look not at sea level itself, but

176   rather at deviations from an ongoing rise in sea level that remain after subtracting off a linear rise over time. Such a subtraction would seem no more meaningful than looking at deviations in temperature rather than at the temperature itself.

Although Singer's analysis is open to question, he is not alone in questioning the IPCC estimates of past and future changes in sea level. In a recent article in *Physics Today* Bruce Douglas and Richard Peltier note that the current rise in sea level is about double what global warming models would predict.[39] This discrepancy could indicate that the models are incorrect, but skeptics have no basis in rejoicing if the IPCC models are underestimating sea level rise.

Some coastal cities and countries in the developed world, such as the Netherlands, have coped successfully with rising sea levels for centuries by erecting dikes. If the sea level rise were to be near the upper end of the IPCC projected range by 2100, then low-lying islands and some coastal cities could be severely impacted. Worldwide approximately 100 million people would be at risk given a one-meter rise in sea level. Sea wall defenses would not be feasible for the small islands, which might have to be abandoned, although the number of people impacted in this case would not be huge. The greatest impact of a rising sea level would probably be in delta coastal regions, such as in densely populated Bangladesh, one of the poorest nations on Earth.

Currently, 25 percent of Bangladesh is one meter or less above sea level, and it could have to be abandoned to the sea. On the other hand, skeptics might argue that if the sea level rise were near the upper end of the projected range, it got there *in part* because of the very heavy fossil fuel emissions associated with a scenario involving an economic convergence between the developing world and the West. In that event, nations such as Bangladesh would presumably have caught up to the Western standard of living and they could cope with the cost of erecting sea wall defenses, much as the Dutch have done for centuries.

The preceding analysis of future sea level rise probably is too

177 optimistic. The biggest source of uncertainty in projected sea level rise involves the models themselves rather than emission scenarios. Thus, it is possible that sea level rise could be near the upper end of the projected range even though future emissions are relatively under control. Furthermore, when we assess the global impact of sea level change it is important to consider not only what might occur by 2100, but what may happen subsequent centuries as well. When the Earth's atmosphere and surface warm, it takes a much longer time for the oceans to catch up, due to their large "thermal inertia." (For example, over the last half-century the upper layers of the oceans have warmed only 0.04°C per decade.) As a result of this delay in ocean warming, increases in temperature originating in the next few decades will continue to raise ocean temperatures—and raise sea level though thermal expansion—for the next several centuries.

Agriculture. There are many reasons to expect that a warmer world will be good for agriculture, assuming a moderate degree of warming. First, the variance between summer and winter temperatures is declining, with the average winter temperature currently warming twice as fast as the summer temperature, making for a longer growing season.[40] Additionally, diurnal (day/night) variations are likely to decrease due to global warming, which should be beneficial, since such variations can also be harmful to some plants. In fact, diurnal temperature variation is so important that crop values might be expected to double with only a 25 percent reduction in day/night temperature differences.[41] Third, overall a warmer world is likely to be a wetter world, because higher temperatures cause more evaporation, resulting in a stronger hydrological cycle.

The IPCC estimates a slight increase in precipitation (1 percent) in the coming century, which should also promote plant growth, although this would be offset by possible droughts associated with the drying of continental interiors. Fourth, new areas of the globe might be available for farming in a warmer world or experience longer growing seasons than are now the

178 case. (Satellite observations during the 1980s showed that high-latitude regions were greening up a week earlier than they had a decade earlier.[42]) Finally, CO_2 has been found to directly promote plant growth, at least in the more than 95 percent of plants of the so-called C3 variety. (C3 plants are those that use a special enzyme in the process of "fixing" carbon to make plant material.) It has been estimated that in the last half-century higher CO_2 levels have increased agricultural output by between 8 and 12 percent.[43]

A particularly interesting aspect of the benefits of higher CO_2 levels in promoting plant growth is that the effects of higher CO_2 and higher temperatures appear to be synergistic. For example, one 1990 study found that when plants were grown in an atmosphere having double the present CO_2 level, the temperature at which they thrived best was 5°C warmer.[44] This combination of more CO_2 and higher temperatures is, of course, just what would exist in a greenhouse warmed world.

A final point about agriculture concerns adaptation. Farmers, despite their small profit margin, have proven in the past that they are adaptable to changing environmental conditions. A gradual increase in temperature could best be coped with if farmers continue to adapt, perhaps by changing their choice of crops. In many cases, this might entail greater dollar yields if higher cash crops, such as citrus (well suited to hot weather), replace less expensive crops. Economists estimate that in the United States a modest temperature increase in the range of several Celsius degrees could significantly improve net cash crop yields, even while the amount of crop land under cultivation might drop.[45]

Surprisingly, even with a temperature increase as large as 5°C, it has been estimated that there would be a net increase in crop yields due to higher precipitation and CO_2 levels. According to economists, the increase could be between 10 and 32 billion dollars annually in the value of U.S. crops if farmers adapt.[46] Farmers in developing nations who are operating on a subsistence basis might not have the luxury of adapting their practices, and they could be harmed by higher temperatures.

179 But in the aggregate worldwide agriculture might benefit from a rise in temperatures, though that might not be true if the rise were at or above the upper end of the IPCC range.

In contrast to the preceding overall scenario for the impact of rising temperatures on world agriculture, the IPCC is much less optimistic. In fact, the IPCC suggests that there will likely be a reduction in crop yields in most regions of the globe if the temperature rise exceeds a few Celsius degrees. In this and some other areas, the IPCC appears to give significantly greater weight to the negative impacts of rising temperatures than the positive ones.

Natural Ecosystems. Many of the same factors that would enhance agricultural production in a warmer world would also benefit natural ecosystems. But there is one crucial difference. A farmer can choose different crops that are better suited to a changed environment, but plants and trees cannot pick up and move to a more suitable climate. Unlike human systems, ecosystems have a much more limited ability to adapt to a large or rapid climate change. Some species, in fact, have a fairly small ecological niche, in terms of the optimum temperature and precipitation level, under which they flourish. Coral reefs are a particularly fragile ecosystem, and many are reportedly in danger due to warmer temperatures plus other causes.

As far as land vegetation is concerned, as global temperatures warmed, we might imagine that various plant species would need to spread their seeds poleward toward cooler temperatures. However, given that many small forest areas are surrounded by developed land, such a poleward migration might not be possible in many cases. On the other hand, it might not be necessary. According to Patrick Michaels, no poleward migration of plant species would be necessary, given the observed synergistic link between higher CO_2 levels and higher temperatures for optimum growth.[47] Michaels is essentially saying that the two factors of rising temperature and CO_2 will offset each other to cause the optimum growth rate for a given species to remain where it is now located. Higher temperatures

180 and CO_2 levels are indeed synergistic, but it is unclear if the two would rise by just the right amounts as Michaels suggests.

The IPCC, on the other hand, is much less sanguine than the skeptics about the vulnerability of natural ecosystems, and notes that some of them may undergo "significant and irreversible damage" because of their limited ability to adapt.[48] Further, the IPCC notes that "while some species may increase in abundance or range, climate change will increase existing risks of extinction of some more vulnerable species." One could read this IPCC comment to say that when climatic conditions change, some species will thrive and others will lose out— exactly as has happened for millions of years before human-caused global warming. Even during the age of the dinosaurs when global temperatures were around 10°C higher than at present, and CO_2 levels ten times higher, the planet was teeming with life. (The preceding observation may have questionable relevance to the impact of a very large temperature change on natural ecosystems occurring in a timescale measured in decades.) Finally, the adverse impact of climate change on natural ecosystems needs to be considered in combination with other stresses, including those associated with land use and various pollutants.

Frequency of Extreme Events. Extreme weather events, such as floods, tornadoes, and storms, can take many lives and be the cause of extensive economic loss. This category could be the most important one of all in terms of the potential impact of global warming. It is estimated that tropical storms have cost an average of 15,000 lives per year worldwide over a 33-year period. Droughts are probably the most costly climatic event of all in terms of lives lost, but the numbers are difficult to quantify, because they occur over an extended period of time and result in deaths involving many other contributing causes.

According to the IPCC, there were "relatively small increases" in areas of the globe experiencing either severe drought or severe wetness during the past century, and in many areas those changes appear to be dominated by natural variability.[49]

181 There also appears to have been no obvious pattern of increase in the intensity or frequency of severe storms. In the case of the United States, virtually every study has shown no trend toward increased droughts or floods.[50] If greenhouse warming is already well underway, the nonoccurrence of a greater frequency of extreme weather events overall is reassuring, even if does not prove that extreme weather events might not increase in frequency if more extensive warming occurs.

Does the IPCC report give a coherent view on the question of extreme events? Despite the cautionary IPCC statements noted previously, the report contains other statements that seem to imply that increases in extreme weather events have already been observed. For example, the IPCC report emphasizes that "there are preliminary indications that some human systems have been affected by recent increases in floods and droughts." That statement could be read to imply that there has, in fact, been an increase in floods and droughts. However, a few sentences later in the report it is noted that the increases have occurred "in some areas," and still later it is noted that risks of drought in the last half-century are likely to have occurred "in a *few* areas." (It would be indeed remarkable if the risks of drought have not increased in some or "a few" areas in any given half-century.)

Another example of an IPCC statement that, if taken in isolation, could give a false impression regarding the likelihood of future extreme events is the projection that "increased summer drying over most mid-latitude continental interiors and associated risks of droughts is likely." Since the IPCC also projects higher precipitation levels in the coming century on a worldwide basis, it is clear that other areas of the globe, especially continental margins, will experience more precipitation and less droughts. Sure enough, at another point in the report it is noted that there is likely to be an increased availability of water in some currently arid regions. (Nowhere does the report indicate whether extreme droughts might be expected to increase or decrease on a worldwide basis in the coming century, or by how much.)

182   Regarding tropical storms, the IPCC report notes that increases in tropical cyclone wind and precipitation intensities are *"likely over some areas."* This assertion naturally leads one to wonder whether they are likely to decrease over other areas, and what the overall trend might be. For storms in midlatitude regions the IPCC projects that we can expect an "increased intensity," but it then honestly adds that there is "little agreement between current models" on the matter.

Finally, the IPCC report addresses itself to truly calamitous events that might occur after the year 2100. It notes that if present trends continue, there could be large-scale and potentially irreversible climate change, leading to the collapse of the Antarctic ice sheets and the disruption of the ocean circulation pattern. However, the report then candidly and appropriately admits that "the likelihood of many of these changes is not well-known, but it is probably very low."

Conclusions

For those keeping score, here's my assessment so far on the four questions raised earlier. As can be seen from my answers below, I believe that the skeptics have the better case (by a close call) only on the last two questions:

1. Has there been a global warming in recent decades?
    Almost certainly, yes. However, that conclusion could be invalidated if satellite data continue to show little warming in the coming decade.
2. Was the warming seen in recent decades anthropogenic?
    Probably. If it continues, we'll probably know for sure by 2010.
3. What is the warming likely to be by the year 2100?
    Probably in the lower half of the IPCC projected range— namely under 3°C.
4. Will rising temperatures be harmful or beneficial?
    They are more likely to be beneficial if the rise is modest. If the rise is near the upper end of the IPCC projected range, impacts

183   *are unlikely to be especially harmful for many developed
nations, although they could be significant for developing
nations.*

The disparate impact of global warming on the developing
nations has received attention in the IPCC report and elsewhere. Citizens of a developed nation are unlikely to become
more concerned about the problem of global warming if its
brunt will be felt not by their children or grandchildren, but
by those of the Third World poor now struggling in a meager
existence. Moreover, Bjorn Lomborg, author of *The Skeptical
Environmentalist,* argues cogently that if the West really wanted
to alleviate Third World suffering it could have a greater impact through direct aid to today's poor than through indirect
aid to their descendants by the mitigation of global warming.
Of course, some forms of aid, such as assistance in developing
clean, high-efficiency power sources could help both today's
poor and their descendants.

For the human species, which has survived over most of its
existence by successfully dealing with immediate threats, it is
difficult to grapple with threats of a long-term global nature.
The question of how much we as a society should worry about
global warming needs to be put in a larger context. That context would consider what actions would need to be taken to
mitigate global warming, how costly they would be, and how
much impact those actions would have on reducing the extent
of the warming. It has not yet been demonstrated clearly that
the costs of mitigation are less than the costs associated with
adapting to higher temperatures. Given my summary assessment above, it seems premature to launch drastic actions, such
as mandatory large cuts in CO_2 emissions, that could have high
economic costs. But certainly actions involving a "no-regrets"
strategy, such as energy conservation would be desirable.

The "no-regrets" strategy refers to actions that would be desirable in their own right, whether or not global warming
proves to be a serious threat. Another example of a no-regrets
action besides energy conservation would be the development

184 of more efficient lighting and automobile engines. Actions in the no-regrets category are free in some cases, and in others would even save money. They also would make the United States less dependent on foreign sources of petroleum and reduce the risk of military confrontation.

There are many other reasons for taking a wait-and-see approach to global warming, rather than implementing drastic actions now. First, as time goes on it will become clearer just how serious the problem is likely to be and which models give the best predictions for the likely temperature rise in the coming century. Second, there is little evidence now that we are in store for irreversible climate change or that "our raft will be swept over the falls" if we stay in the river a few more decades. Third, there are possible technical fixes to the problem that need to be explored, even though some of them may sound like science fiction now.

These technical fixes include removing CO_2 from the exhaust gases after fossil fuels are burned, putting many tiny reflecting mirrors in orbit around the Earth to act as a partial sunshade (!), and adding iron to the surface of the oceans to promote the growth of algae that would absorb CO_2. Obviously, some of these measures could have harmful side effects or may prove unfeasible, but they are at least worthy of serious consideration and experimental testing.

The last of the three suggested measures, fertilizing the oceans with iron, actually has been tested to a degree, and the results to date are not promising.[51] According to one recent experiment involving an 8-kilometer-diameter patch of ocean, there was a reduction of CO_2 following ocean fertilization, but the CO_2 removed from the atmosphere does not appear to have sunk into the deep ocean.[52] Sallie Chisholm, a scientist involved in this research, is rather pessimistic about the wisdom of this effort. According to Chisholm, "Artificial fertilization with iron would probably have unintended side effects, such as deoxygenating the deep ocean and generating greenhouse gases that are more potent than CO_2."[53] Moreover, she notes that "in the long run ocean fertilization is not sustainable. So

185 why start?" Removal of CO_2 during fossil fuel combustion at the power plant is probably the most feasible of the three possibilities suggested, but the current costs would be quite high.

Finally, the issue of how much we should worry about global warming needs to be put in the context of all the other issues that humanity needs to worry about. There may well be threats to the planet about which we worry too little, such as the possibility of being hit by a meteorite from space large enough to destroy most life on Earth. Worrying about everything that could possibly happen can be crippling. By declining to worry too much about global warming, we would free up some of that finite amount of "worry time" to more pressing threats. This assessment probably is not too far off from that of the American public, who, while claiming to worry somewhat or a great deal about the issue of global warming, nevertheless ranked it next to last in a list of 13 environmental threats, as indicated in table 6.1.

Exponential Population Growth—The Mother of All Threats?

Any quantity that increases by some fixed percentage P each year grows exponentially in time. Many people believe that modest percentage growth rates such as 1–2 percent do not represent a serious long-term problem. They are mistaken. If world population, for example, were to continue to grow at its present rate of about 1.5 percent per year it would double in 47 years, and double again every 47 years. (In general, the rule is that for any percentage P the doubling time is $70/P$ years.)

At present there is about a million square feet of the Earth's surface for every human being. If everyone were spread out uniformly over the entire planet—land and sea—we would all be a comfortable thousand feet from our nearest neighbor. Were exponential growth able to continue for 20 more doublings, on the average there would be one person per square foot, and we'd all be packed shoulder to shoulder over the entire planet. Clearly, the human population will cease its explosive exponential growth long before that point is reached. The main issue is whether the expo-

186 nential growth is halted by relatively benign means or by increasing death rates due to starvation, disease, war, infanticide, and environmental degradation.

Controlling world population growth is an even thornier international issue than controlling CO_2 emissions. It is understandable that the IPCC would simply assume that world population will somehow stabilize in making its emissions scenarios for the coming century. However, if we are serious about protecting the environment, it is equally short-sighted to fret about global warming (while assuming that world population will somehow stabilize by itself) as it is to worry about population growth while assuming that the climate problem will take care of itself.

Finally, there are some arguments on the other side of the ledger. Although a large amount of global warming by 2100, i.e., at the upper end of the IPCC estimate is probably an unlikely event, it is not impossible. Similarly, an irreversible climate change of the sort considered by the IPCC is not out of the question. Scientists believe that a "runaway" greenhouse effect transformed the surface of Venus to the hellish present point where lead would melt. Can we be absolutely certain that such a runaway greenhouse effect could not happen here? For example, were the polar ice sheets to melt it could be a significant source of positive feedback that would create still more warming than expected.

A Runaway Greenhouse Effect?

A runaway greenhouse effect is more than a steadily progressive warming as greenhouse gases continue to be added to the atmosphere by humans. The two distinguishing characteristics of a runaway effect are the positive feedback that drives it even if anthropogenic emissions should stabilize, and the irreversibility of the process. Here's how the process is believed to have occurred on Venus. Venus has water just as Earth does,

187 but, unlike on Earth, any initially liquid oceans on Venus would have steadily boiled away or converted to vapor. That's because on Venus as water vapor got added to the atmosphere due to greenhouse warming, it would cause still more water to evaporate because of the higher temperature, until eventually all the liquid water was gone. The same positive feedback also operates on Earth, but only to a degree. That's because the Earth is further from the sun than Venus, and receives only about half the solar radiation as Venus. As a result, on Earth the positive feedback due to water vapor would be self-limiting. At some point the atmosphere would become saturated with water vapor and no further evaporation would occur. According to climatologist John Houghton, "There is no possibility of such runaway greenhouse conditions [as occurred on Venus] occurring on the Earth."[54]

Wild temperature excursions in the opposite (cooling) direction are also not impossible. For example, freshwater from melting ice in the northern Atlantic could disrupt the gulf stream and be the trigger for a new ice age, as may have happened in the past. (Not that long ago many scientists were concerned about a new ice age, rather than global warming.) It does seem prudent to pay some attention to very low probability events that could have very great consequences. In addition, we need to pay attention to long-term climate changes extending well beyond 2100, especially sea level rises that could continue for several centuries due to today's warming.

All things considered, I'd rate the idea that we shouldn't worry too much now about global warming at 1 flake. But the matter does call for watchful waiting. It could warrant urgent attention in the coming decades if the trend in temperatures were to lie in the upper half of the IPCC projections.

"Ahhh, this porridge is just right," she said happily as she ate it
all up.

—*Goldilocks*

JUDGING BY THE POPULAR MEDIA, we live in an era in which
alien beings are everywhere. Films, TV shows, and the popular
press are replete with stories about both friendly and not so
friendly visitors from space. According to a 1996 Gallup Poll, 45
percent of Americans believe that aliens have visited Earth, and
an astounding 71 percent believe that the government is covering up its knowledge that aliens exist.[1] (Taking these two results at face value, it would seem that 26 percent of Americans
must believe the government is covering up merely the discovery of alien life elsewhere, not its presence on Earth.)

I have never understood why the government might want to
cover up such a discovery, even if it could. The usual explanation—that the government does not want to cause a panic—
makes no sense if most people believe aliens exist anyway. I
assume the supposed cover-up is based on the government
having been taken over by the aliens masquerading as
humans—something I've long suspected since reading a
tabloid report about six particular U.S. senators being space
aliens (just the six I would have guessed too!).

Most astronomers consider beliefs in alien visitations and
government cover-ups to be complete nonsense, although they
are much more accepting of the possibility that alien civilizations exist elsewhere in the universe. Some scientists, like the
late Carl Sagan, are optimistic, believing that as many as a million or more intelligent civilizations might exist in our galaxy
alone.[2] Other scientists, like the late physicist Enrico Fermi, are

189 more cautious. Fermi asked simply: "If they exist, where are they?"[3] He reasoned that surely some advanced civilizations would eventually master interstellar travel, and so their absence from the scene could only mean one of the following:

1. Intelligent aliens don't exist.
2. Interstellar travel is impossible.
3. All extraterrestrial civilizations have decided that space travel is not worth the effort.
4. Intelligent civilizations don't last long enough to reach the point where they can make the trip.[4]

A pair of scientists, geologist Peter Ward and astronomer Donald Brownlee, recently have come down on Fermi's side of the debate. In their book *Rare Earth: Why Complex Life Is Uncommon in the Universe*, they have argued that, while simple life is probably extremely common throughout the universe, complex life—even life at the level of green slime—is very rare.[5] The idea is that Earth has such a restrictive set of environmental conditions and planetary history that complex life is unlikely to be found anywhere else. In this chapter, we'll first outline the basis for the "rare earth hypothesis" (REH), and then critique it.

How Did Life Originate on Earth?

Various theories have been advanced as to how the simplest living system might have evolved from a prebiotic "soup" of organic chemicals. One idea is that a set of "autocatalytic reactions" might have occurred, in which reacting chemicals stimulated one another's creation—life in effect pulling itself up by its own bootstraps. Such autocatalytic or "self-replicating" reactions are known to exist in nature, as in the case of a 32-amino acid peptide, which is capable of self-replication, as long as outside "food" sources exist.[6] However, short of successfully creating life in a test tube, such theories of life's origin must necessarily remain speculative.

190   Nevertheless, because this first step in creating life involves some unknown mysterious process, most people believe that the initial life creation event is the "hard" part—a view not shared by many planetary scientists. Once life was created on Earth and evolution had something to work with, it would seem only natural that more and more complex forms would arise given enough time. Thus, the "easy" part would be the evolution of complex life arising in a series of small steps from simple life. The rare Earth hypothesis says that conventional wisdom has it exactly backward: the easy part is the origin of simple life (which could be widespread), and the hard part is the later development of complex life (which is probably very rare in the universe).

Ward and Brownlee's REH is a "crazy" idea because it goes against some people's conventional wisdom, though it probably fits quite comfortably alongside the views of many religious people, who would be more likely to view the Earth as a very special place. As already noted, the first half of the REH says that simple life is probably widespread throughout the universe. By simple life we mean something akin to bacterial life, which consists of single cells.

The evidence for the idea that something like bacterial life is widespread in the universe is indirect, since the only examples of life of any kind we have is life on Earth. Space probes sent to the moon and Mars have not detected life, and probes have yet to be sent to search for life elsewhere in the Solar System. (Later we'll consider possible evidence for *ancient* Martian life.) In searching for life beyond the Solar System, given the difficulties of interstellar travel, we may need to rely entirely on Earth-based observations rather than space probes, at least for a very long time.

Why Is Primitive Life Thought to Be Common in the Universe?

The indirect evidence of primitive life in the universe is of two types. First, primitive life seems to have arisen on Earth at the

191 earliest date it possibly could have. One sign of very early life
on Earth are the oldest fossils, the so-called stromatolites that
arose 3.6 billion years ago. An even earlier signature of life is
the effect it had on the Earth's atmosphere. Each chemical element in nature comes in different isotopes, which are defined
by the number of neutrons contained in the atomic nucleus.
Most chemical reactions do not distinguish between isotopes of
an element, but biological processes often do. During photosynthesis, for example, the lighter isotope carbon-12 is more
readily absorbed than the heavier isotope carbon-13. On this
basis, scientists have inferred that primitive photosynthetic life
was present 3.8 billion years ago by looking at the carbon isotope ratio of ancient Greenland rocks formed around that time.
That early date would place the first primitive life about 0.7 billion years after the Earth first formed from the cloud of gas and
dust swirling around the newly formed sun.

However, no life could survive on the surface of the Earth
during its first half-billion years, given the continual rain of
death from the skies. Even now, large comets and asteroids occasionally bombard our planet. But during the early days following the planet's formation, the amount of space debris—
and hence the frequency of impacts of large bodies with
Earth—was far greater than now. (Over time, space debris
would tend to be cleared away by falling toward planets and
being absorbed.) It has been suggested that even if life somehow did arise on Earth during the first half-billion years, it
probably would have been wiped out during this period of
continual heavy bombardment from space. So, 4.0 billion years
ago is just about the soonest life could have arisen and per-
sisted. Finding evidence of life this for back suggests that primitive life arises anywhere "easily and quickly," assuming the
raw materials are present.

The second basis for believing that primitive life may be
widespread throughout the universe are findings of primitive
life on Earth under far harsher environments than once were
believed possible. Bacteria and bacteria-like organisms (known
as extremophiles, or lovers of extreme conditions) have been

192 found near undersea thermal vents existing at temperatures as high as 167°C. These organisms have also been found living in cracks in solid rock down to depths of 3.5 kilometers. In the latter case there is a complete absence of light—making photosynthesis impossible—and very little liquid water present. These simple organisms are believed to belong to a domain known as archaea, and they may be older than the first bacteria. Thomas Gold, a Cornell University scientist, has suggested that such subsurface life may even be Earth's most massive life form, with a total mass that equals or exceeds all surface life.[7]

The harshness of the environments where extremophiles are found suggests that they could exist on other planets or elsewhere throughout the universe. They could even exist inside comets, meteorites, and interstellar grains of dust that drift through space. Life on Earth might well have been "seeded" by such bodies during or shortly after the period of heavy bombardment during its first half-billion years. Underground extremophiles might even have survived the latter stages of the bombardment in their deep underground "bomb shelter," and then seeded life on the surface after the bombardment became less severe.

Panspermia: The Seeding of Life from Space

The presence of primitive life on Earth so soon after conditions would have permitted it to survive suggests the possibility that life might have been seeded from space (figure 7.1). This idea, known as panspermia, has a long history. In the early 1900s, for example, the Swedish chemist Svante Arrhenius suggested that bacterial spores could be transported to Earth from space by the pressure of sunlight.[8] The idea of panspermia was revived in the 1970s by Fred Hoyle and Chandra Wickramasinghe.[9] Their observations of the spectrum of light passing through interstellar dust supported the presence of bacterial spores in the dust, which continues to reach Earth at a rate of one grain per square meter each day. Many scientists today are skeptical of panspermia, however, because they believe that life could not survive the radiation present in space on its

Figure 7.1 "Honey, do you remember that movie *The Invasion of the Body Snatchers*?"

long journey to Earth. On the other hand, there are various ways around that objection. Bacterial life present inside comets would be sufficiently shielded from radiation and could be spread along with the comets throughout the galaxy. As far-fetched, as it might sound, we cannot rule out the possibility that intelligent aliens have seeded primitive life on Earth billions of years ago—perhaps as a high school science project!?

Apart from the possibility of panspermia, the idea that primitive life may be widespread throughout the universe is given support by its extreme hardiness and its very early appearance on our planet. Furthermore, the idea that life evolved from nonliving matter also seems fairly plausible to many people.

194 (These ideas are not intrinsically in conflict with a belief in God, although they do conflict with some religious beliefs.) But it is surely the second half of the rare Earth hypothesis that is the "crazy" (or, at least, very controversial) part—namely the idea that complex life is very rare in the universe. On the other hand, many religious readers may not find this idea crazy at all, and be quite accepting of the rarity or uniqueness of advanced earthly life—a tradition going back to the belief that Earth occupies a special place in the cosmos.)

Why Complex Life in the Universe May Be Rare

"Complex" life in the context of the REH not only refers to the higher species exhibiting some level of intelligence, but includes the simplest sorts of animals and creatures, such as sponges. Of course, animal-like creatures on another planet might not have much in common with earthly animals, so it is difficult to be certain what the range of conditions might make such evolution possible. But, we can draw some tentative conclusions from what we observe on Earth. Let us briefly sketch out the conditions required for complex life according to Ward and Brownlee, and explain why they think it is so rare. Later we will try to evaluate the plausibility of that view and see how it could be put to the test.

One condition usually assumed necessary for the presence of complex life is the presence of liquid water, although it is conceivable that on another planet a different liquid, such as ammonia, might play an equivalent role. Let's stick with water, however, as being an essential ingredient for "life-as-we-know-it." Water remains a liquid for the temperature range 0 to 100°C at one atmosphere pressure. But that range may both understate and overstate what is necessary for complex life to evolve. On Earth animals cannot exist at temperatures higher than around 50°C, which may be the upper limit, based on the chemical stability of the cell membrane.[10] (All animals, both single-celled amoebas and multicellular higher forms are composed of

195 "eukaryotic" cells. These cells have nuclei that are more complex than the more primitive "prokaryotic" cells of most bacteria.) On the other hand, the 50°C limit may be incorrect, because water can remain a liquid at temperatures much higher than 100°C, provided the pressure exceeds one atmosphere.

Many of the conditions needed to allow animal life to evolve and survive on a planet relate to the need to maintain temperatures in some habitable range. Table 7.1 shows a list of all the conditions as outlined by Ward and Brownlee.[11]

Not Too Hot and Not Too Cold. Some of the conditions listed in table 7.1 are fairly obvious. For example, the idea that a planet needs to be the right distance from its star for complex life to exist is just a matter of keeping planetary surface temperatures nice and comfy—inside the so-called "habitable zone." Given a star similar to our sun, a planet could not have liquid water on its surface if its distance to the star were much different than the actual Earth–sun distance. The exact inner and outer boundaries of the habitable zone in the Solar System are a matter of debate, but Earth's two closest neighbors, Mars and Venus, both appear to be outside of it. Venus, which is 30 percent closer to the sun than Earth, has surface temperatures hot

Table 7.1

What Is Required for a Planet to Be "Habitable"?

Conditions relating to the planet itself		
Mass	Distance to star	Axis tilt
Atmosphere	Amount of carbon	Amount of water
Magnetic field	Plate tectonics	Large moon
Snowball Earth	Inertial Interchange Event	
Conditions relating to planetary system, star, and galaxy		
Type of star	Stable planetary orbits	Jupiter-like neighbor
Type of galaxy	Mars-like neighbor	Position in galaxy

196   enough to melt lead. If Venus ever did have surface water it has long ago boiled away. On Mars, which has a smaller mass and less gravity than Earth, the air pressure is too low to keep water from boiling away. (At one time, when its atmosphere was denser, Mars probably did have liquid water, but it may be now frozen in the Martian crust.)

The existence and the location of habitable zones around other stars depend primarily on each star's temperature. A star's temperature, in turn, depends on its mass, since more massive stars burn hotter than the sun. Planets orbiting stars that are heavier (and hotter) than our sun would need to have orbit radii greater than the Earth–sun separation to retain surface water. However, it is unclear if stars with masses appreciably greater than that of our sun could even have a habitable zone.

Fast Burners Don't Last. Why are massive stars so unpromising? Massive stars burn much hotter than our sun, and consequently have much shorter lives, because they exhaust their fuel more rapidly. Our sun will probably enter its so-called "red giant" phase in another 4 or 5 billion years, at which point it will balloon outward and swallow up the inner planets, extinguishing any life on Earth. A star with a mass 50 percent greater than the sun would enter its red giant phase after it was only 2 or 3 billion years old. Two billion years might not be enough time for complex life to evolve on a planet, given that the first signs of complex life on Earth appeared when it was 2.5 billion years old. Another reason massive very hot stars are unpromising places for complex life is that they emit much greater quantities of dangerous ultraviolet radiation than the sun.

A Reminder

The preceding discussion laid out Ward and Brownlee's case that massive stars are an unpromising place for the development of complex life. In point of fact, this and other requirements for complex life may not be

197 nearly as restrictive as Ward and Brownlee seem to believe. But in keeping with the spirit of our discussion so far, we'll delay a critical look at their REH until after laying out their case in its entirety.

Cool Stars Make You Get Too Close. If stars much more massive than our sun are unpromising, what about stars that are less massive and cooler than the sun? These stars are much more common in the galaxy than our sun. Planets orbiting a less massive (and hence cooler) star would need to be closer to it to have Earth-like temperatures, just as you'd need to be closer to a small campfire than a large one to remain comfortable on a cold night. However, in most cases, the smaller distance to a cooler star would cause the planet to experience "tidal lock," meaning that it always keeps the same face toward the star, just as the moon always keeps the same side facing Earth. A planet experiencing tidal lock would have extremely hot temperatures on the side always facing the sun and frigid temperatures on the opposite side, making it poor candidate for the evolution of complex life over most of its surface. Thus, only a very limited percentage of all stars are likely to have habitable zones, defined in terms of a suitable temperature for surface liquid water on a planet.

Two Is Not Better Than One. The conditions for complex animal life get even more restrictive when we consider other properties of stars besides their mass (and temperature). For example, a majority of stars appear to be binary or multiple stars—two or more stars in orbit about each other. It is unlikely for stable planetary orbits to exist in such cases, since a planet orbiting one of the stars (or the two combined) is likely over time to have its orbit disturbed, and possibly even find itself flung into interstellar space.

Location, Location, Location. The location of a star can be crucial in determining whether complex life could evolve. For example, stars that are nearer to the center of the galaxy than our

198 sun tend to be much closer together. As a result, the chances of a collision with another star (very remote in our region of the galaxy) become significantly higher. More importantly, the chances of receiving a lethal blast of radiation, when a neighboring star undergoes a supernova explosion, also become significantly greater. Perhaps the central 25 percent of the galaxy can be excluded as potential homes for the evolution of complex life on this basis.

The outer reaches of our galaxy are also an unlikely place for complex life to evolve for a very different reason. Many stars in our region of the galaxy tend to be enriched in elements heavier than helium, which were cooked up in earlier generations of stars and expelled when the stars underwent supernova explosions. In contrast, stars in the outer regions of the galaxy tend to be mostly older "first-generation" stars, lacking all the elements heavier than helium that are essential to life.[12] Similarly, we can exclude not only many sections of our galaxy, but also other entire galaxies as places complex life could evolve on exactly the same basis. Unlike our spiral-shaped Milky Way Galaxy, many other galaxies having an elliptical shape contain nearly all older (first-generation) stars, which, as already noted, lack the heavy-element building blocks for life.

Not Too Big, Not Too Small. Many of the conditions necessary for complex life to evolve and survive relate to details of a planet itself. Not surprisingly, the mass of the planet is important, since if it were too small—comparable in size to Earth's moon—it would not be able to retain an atmosphere unless it was far from its parent sun. Conversely, if the planet has too large a mass it could have much too dense an atmosphere. A planet with a very dense atmosphere is likely to experience a "runaway greenhouse effect," as appears to have been the fate of Venus. The initial composition of a planet's atmosphere is also important if animal life is to eventually evolve. The Earth is believed to have had an atmosphere initially rich in CO_2.

199 This atmosphere later made photosynthesis possible, which was a precursor to the evolution of an oxygen atmosphere, a necessary condition for animal life of an earthly variety.

Some Surprising Conditions Needed for Complex Life

Many of the entries in table 7.1 seem surprising. For example, why should the planet need to be in a system that includes Mars and Jupiter clones not too far away from it? It is now known that the Earth has repeatedly undergone a series of ancient "mass extinctions," many of which were caused by meteorite and comet bombardment. One such mass extinction occurred about 65 million years ago, when the dinosaurs became extinct, along with a majority of the species then existing on Earth.

"Good" and "Bad" Jupiter Clones. Jupiter serves as Earth's "space-debris removal system." A large Jupiter-like planet serves to sweep away from the inner Solar System many comets and asteroids that might otherwise strike our planet. It has been suggested that the frequency of catastrophic impacts of 10-kilometer-diameter bodies with Earth might be 10,000 times greater if Jupiter were not present in the Solar System. In that case, instead of a planetwide catastrophe every 100 million years or so, we might experience one every 10,000 years, which would probably make it impossible for complex life to evolve and survive.

A Jupiter clone can play this life-saving role only if, like our "good" Jupiter, its orbit is beyond Earth and is not highly eccentric (noncircular). If Jupiter were in a highly eccentric orbit or in an orbit closer to the sun, it would be likely to disturb Earth's orbit, possibly even causing Earth to be ejected from the Solar System! Such "bad" Jupiters could be catastrophic for the development of complex life on Earth. (In searching for planets around other stars, astronomers have observed many stars orbited by bad Jupiters.[13]

200 *Hitchhikers from Mars?* The claimed need for a Mars-like clone in the planetary system is even stranger than that for a Jupiter-clone. Ward and Brownlee suggest that primitive life might have begun not after, but rather *during* the time of heavy bombardment in Earth's early history, but that it was repeatedly wiped out on Earth due to the sterilizing impacts of large comets and meteorites. Mars, with its thin atmosphere and low escape velocity, would be a good place from which primitive life could "hitchhike" a ride on a meteorite. Hypothetically, if life originated on Mars, some rocks blasted off the Martian surface by meteorite impact might then reach Earth, where they could seed earthly life, even after it had been repeatedly extinguished. Thus, *assuming* that the period of heavy bombardment was somehow necessary for life to evolve on both planets, a Mars-like neighbor serving as a "refuge" could reseed earthly life after it had been repeatedly extinguished.

A Cosmic Ray Shield. Another less-than-obvious Ward and Brownlee condition for complex life to survive on a planet is the existence of a magnetic field. Earth is unique among the planets in having a large iron core, which creates a sizable magnetic field. The Earth's magnetic field causes the paths of charged particles like protons and electrons to curve, and therefore it deflects charged particles among the cosmic rays bombarding the Earth. As a result, living organisms on the surface of the Earth experience much lower levels of cosmic radiation than would be the case if the Earth had no magnetic field. Radiation levels might be lethal to complex life on Earth were there no magnetic field.

Just the Right Axis Tilt. One more surprising condition relates to the tilt of the Earth's axis with respect to a line perpendicular to the plane of its orbit. The Earth's axis tilts by around 23.5 degrees, which is primarily responsible for the climate changes occurring during the course of the year. Without any tilt to its axis, Earth's climate would be nearly the same the entire year, apart from a small change due to the varying Earth–sun dis-

201 tance. Conversely, if the Earth's tilt were much greater than its actual value, the climate change during the year would become much greater—possibly even extreme enough to jeopardize the survival of animal life. A similar fate might occur if the tilt of the Earth were to undergo random changes over time. Such random variations are predicted to occur in the case of a planet lacking a large moon like our own. In fact, it has been estimated that were it not for our stabilizing moon, Earth might flip over on its side (a tilt of 90 degrees) in less than tens of million years, causing catastrophic changes in climate. (Given the unusual circumstances believed to surround the origin of the Earth's moon, such large moons are believed to be rare for Earth-like planets.[14])

Shake, Rattle, and Roll. One particularly surprising condition for the evolution of complex life on a planet cited by Ward and Brownlee is the existence of plate tectonics, which is said to be unique to Earth among the planets in the Solar System. The Earth's crust and upper mantle consist of slowly moving plates that slide over the underlying mantle. These plates cause mountains to rise at places where one plate goes beneath another, and they cause earthquakes to occur when one plate slips and sticks as it rubs sideways against another.

The motion of the plates also acts to recycle material between the atmosphere and surface of the Earth and the deeper layers, since new material comes up from below as plates separate. Plate tectonics is, therefore, considered important to the evolution of complex life on Earth because it recycles elements, particularly carbon and mineral nutrients needed by plants, and therefore also acts as a "global thermostat." The way this works is that during times when it is warmer than usual (and more rain than usual occurs), more weathering of rocks occurs and more CO_2 than usual is removed from the atmosphere. Since the same amount of carbon continues to be brought up from below by volcanoes largely caused by plate tectonics, the net result is to remove some net amount of CO_2 from the atmosphere, and cause less greenhouse warming, or act to cool the planet.

202   Plate tectonics is also important to complex life because it is responsible for the gradual increase in land area over time. Without plate tectonics and the building of mountain chains, the Earth's surface would have a much more uniform elevation. The result would be shallow oceans covering a much more watery world. Thus, plate tectonics has greatly increased the number and diversity of habitats on which animals could evolve. In addition, the development of an oxygen atmosphere may have been due partly to plate tectonics, which recycles nutrients and stimulates photosynthesis. Given all the factors associated with plate tectonics, it apparently has been a crucial factor in the development of animals on Earth. According to Ward and Brownlee, "It may be that plate tectonics is the central requirement for life on a planet, and that it is necessary for keeping a world supplied with water (through control of the thermostat)." Moreover, they note that "plate tectonics is found in our Solar System only on Earth, and it may be "vanishingly rare in the universe as a whole.[15]"

Snowball Earth. This is perhaps the strangest condition that Ward and Brownlee suggest may be necessary for complex life to evolve on a planet. The Earth has gone through a series of ice ages during which glaciers advanced and global temperatures dropped appreciably. The granddaddy of all ice ages—called "snowball Earth"—is believed by some geologists to have occurred on at least two occasions, once around 2.5 billion years ago, and more recently around 600 million years ago.[16] Unlike other ice ages, glaciers during the two snowball Earth episodes extended to all latitudes, including the equator. Even the oceans may have been covered with ice, at least during the first episode. The evidence for snowball Earth—deserving of a "crazy idea" label all its own—is the finding of small angular rock fragments and so-called drop stones due to the motion of glaciers at all latitudes around the globe.

The idea that climate instabilities could cause the Earth to experience a runaway "ice house effect" is the counterpart to the concept of a runaway greenhouse effect. If, for example, global

203 temperatures should drop and glaciers should expand, the resulting whiter Earth would reflect more of the sunlight falling on it back to space. Increased reflection means less absorption of sunlight and still colder temperatures, and thus a vicious cycle, making for a colder and colder world. Without something to counteract this trend, more and more of the planet's surface would become ice covered. It used to be thought that if the Earth ever did become completely ice covered, the planet could never escape the death grip of the ice, but, obviously, that has not been the case, assuming the snowball Earth theory is correct.

Volcanoes to the Rescue. The rescue from the snowball Earth episodes was probably provided by volcanoes. When volcanoes belched forth quantities of carbon dioxide, the gas slowly accumulated in the atmosphere, as there was no photosynthetic or geological processes to remove it. Gradually, the increasing levels of atmospheric CO_2 gave rise to an increasing greenhouse effect, which melted the ice. The melting of the ice would have exposed to the air much iron and long-covered nutrients. The result would have been spectacular, namely a massive amount of photosynthesis and the evolution of an oxygen atmosphere. The preceding scenario may help explain why the occurrence of Earth's release from its two snowball ice episodes coincided in time with two events that were crucial in the development of animal life on Earth.

The most recent of these two events was the "Cambrian explosion"—the proliferation of all sorts of life forms not previously seen on Earth, which occurred around 540 million years ago. The date is based on the sudden appearance in the fossil record of the remains of complex multicellular organisms, with virtually nothing seen at earlier times. The first complex multicellular organisms were very tiny animals, such as flat worms, which escaped the geologists' notice initially in the pre-Cambrian fossil record. It has been suggested that the last snowball Earth episode, which nearly put an end to all life on Earth, ironically was the stimulus that later created this proliferation of complex life forms.

204   A similar great advance in the journey toward complex life may have been associated with the first snowball Earth episode. It is believed that the eukaryotic cell found in all complex life forms first evolved shortly after the first snowball Earth episode. The eukaryotic cell is much more complex than the prior simple (procaryotic) cell of bacteria, having roughly a thousand times the length of genetic code and a thousand times the size. This new type of cell—and the concurrent invention of sexual reproduction—is what then made advanced multicellular organisms possible. Sex is also important. Its "invention" made more rapid evolution possible, because new organisms could form whose genetic code is a recombined scrambled version of that of the two parents.

Evidence for the date of the origin of euklaryotic cells is not based on fossils. Rather, the evidence is based on dating that relies on DNA mutations occurring at some known rate over time. The idea is that if genetic mutations occur at a known constant rate in time, we can tell whether all the organisms in a given family have descended from a given common ancestor, based on the amount by which their genes differ from one another, and then extrapolating backward. The important lesson here is that complex life on Earth appears to have advanced not in a series of small steps but mainly in two great jumps. Each jump was preceded by a snowball Earth episode, which effectively gave the message to life then existing: "evolve or die." Snowball episodes may, therefore, be one more of a long list of conditions that need to occur on a planet if complex life is to evolve and survive—or so claim Ward and Brownlee. And now for the critique.

Are the Previously Discussed Conditions for Complex Life Valid?

The second part of the rare Earth hypothesis rests on the claim that the conditions for complex life to evolve are much more restrictive than for simple life (see table 7.1 for a summary). Here we shall examine this part of Ward and Brownlee's claim and

205 see how well it stands up to critical scrutiny. Some conditions listed in table 7.1 apply equally to simple and complex life. The notions, for example, that complex life could not evolve in star systems lacking elements heavier than helium or that it could not evolve near the center of the galaxy are probably true, but the same could be said about primitive life. If we want to show that primitive life is abundant and complex life is rare, we need to focus on conditions that apply *only* to complex life.

What about those conditions relating to the habitable zone around a star, based on the need for liquid water to exist on the surface of a planet? Defining a habitable zone so narrowly that both Mars and Venus in our own Solar System would lie outside it seems unduly restrictive. Mars, for example, did apparently have liquid water on its surface at one time, based on the appearance of dried up river beds (see figure 7.2). The planet might still have liquid water if it were more massive and had

Figure 7.2 A canyon on the surface of Mars that may have been formed by flowing water present at some time in the past. The area shown is about six miles across. Picture courtesy of NASA/GSFC.

206 enough gravity to retain a much thicker atmosphere, where a sizable greenhouse effect could operate.

And on what basis do Ward and Brownlee require that liquid water must be on the *surface* of a planet for the evolution of complex life? Europa, one of Jupiter's moons, is believed to have a subsurface ocean. Since complex life probably evolved on Earth in the ocean, we cannot rule out a similar scenario for Europa. On this basis, the animal habitable zone might well extend throughout the Solar System. (In fact, the moons of Jupiter-like planets might be the "garden spots" in the universe.)

Similarly, it may be unduly pessimistic to conclude that the large majority of stars dimmer than our sun are poor candidates for complex life. Recall that for such stars the habitable zone would have to be so close to the star that a planet would experience tidal lock, with one side roasting and the other freezing. However, that scenario wouldn't apply in the case of a planet massive enough to retain a thick atmosphere. Such a planet could be at a much greater distance from the sun, because an extensive greenhouse effect could compensate for the lower level of incoming solar radiation and keep the surface temperatures in a comfortable zone. In like manner, it is also too pessimistic to exclude all binary stars as promising candidates for the evolution of complex life. According to Wolfgang Brandner, stable orbits of planets are possible in about a third of all binary systems, specifically systems in which the two orbiting stars are either closer together than the Earth and Sun or at least 400 times further apart.[17]

The need for a planet to have a magnetic field for complex life to evolve is also highly questionable. It is true that the Earth's magnetic field does deflect some cosmic rays and reduce the radiation level on the planet's surface. However, the shielding effect of the atmosphere contributes a *much* greater reduction in cosmic ray intensity than the magnetic field. (This fact can be confirmed by noting that the radiation levels at the north and south poles, where the magnetic field has little effect on deflecting cosmic rays, is not much different than radiation

207 levels at the equator.) A planet lacking a magnetic field would experience the same radiation levels as Earth if its atmosphere were only a mere 10 percent thicker.[18]

The presence of a large moon was listed as a condition for complex life to evolve, based on the moon's stabilizing effect on the tilt of the planet's axis, and hence climate stability. Our own moon is believed to have resulted from the impact of a very large (Mars-sized) body with Earth early in the planet's history. Computer simulations show that only for a restrictive set of conditions would one large moon have formed as a result of such a collision. But does that fact really mean that large moons are necessarily rare in other planetary systems? The collision route to formation of a moon is only one of several ways a large moon could form, and the fact that our moon may have formed under "unusual" circumstances shouldn't imply that large moons are necessarily rare in other planetary systems. Even restricting ourselves to our own Solar System, we do find one other planet (Pluto) with a moon having a comparable size to the planet itself. So, if we exclude the so-called Jovian planets (Jupiter, Saturn, Uranus, and Neptune), which have many small moons, planets with large moons are two out of five, which is not particularly rare.[19]

And what of the claimed need for a planet to have plate tectonics for complex life to evolve and survive? Recall that plate tectonics was said to have many benefits, including the recycling of materials (especially carbon and oxygen), so as to provide a planetary thermostat, the creation of continents (so as to promote biodiversity), and the creation of a magnetic field. As we shall see, the first two claimed benefits of plate tectonics may be as unimportant for the evolution of complex life as a magnetic field was shown to be.

The planetary thermostat can be regulated by a variety of mechanisms apart from plate tectonics. Furthermore, the "regulation" provided by a nonbiological means such as plate tectonics is not guaranteed to keep atmospheric oxygen and CO_2 levels in a range most conducive to the evolution of complex life. Hartman and McKay have gone so far as to claim that plate tectonics is actually a *hindrance* to the development of complex life, by its action in slowing the recycling of carbon and oxygen.[20] They note that Mars, which lacks plate tectonics, may have actually seen complex life evolve within 100 million years of the planet's formation—25 times sooner than Earth—due to the more rapid oxygenation of the Martian atmosphere. It may not be ample time that is the limiting factor in the evolution of complex life, but rather the development of an oxygen atmosphere.

But even if this speculation is wrong and plate tectonics really is essential to the development of complex life, we can ask under what conditions would tectonics occur. According to planetary geologist V. Solomatov, the one crucial factor for a planet to have plate tectonics is the presence of liquid water.[21] In that case, the rarity of plate tectonics in the Solar System simply reflects the rarity of surface water, and it would not be an *"extra"* condition for the emergence of complex life on a planet.

Some astronomers believe that comets may have brought much of Earth's supply of water. If that hypothesis is true, it is possible that *any* planet having enough gravity and the proper range of temperatures would be very likely to have liquid water on its surface. One indication that liquid water may not be rare in other planetary systems would be the detection of water in the atmosphere of another star or, more likely, a planet associated with it. Recently, NASA scientists have used the Hubble Space Telescope to detect the atmosphere of a planet around another star (HD 209458).[22] They hope eventually to find evidence for gases such as water vapor and methane that could suggest the possibility of life.

Evolving animal life on a planet is one thing, but having it survive and flourish is quite another. The claimed need for a planetary system to have a Jupiter clone to permit complex life to survive was based on the effect that such a large planet would have in reducing the number of large comets and meteorites that might otherwise strike a terrestrial planet. Searches made for extrasolar planets have shown that about 5 percent of

209   the systems studied so far do, in fact, have planets with masses comparable to Jupiter—but mostly "bad" Jupiters.[23]

It is not yet possible to say what fraction of planetary systems have "good" Jupiters, since present detection methods make it difficult to find large planets unless they are close to their star—although one such system was discovered in 2002 around a star in the constellation Cancer (see figure 7.3).[24] Nevertheless, as Ward and Brownlee note, "It is possible that up to 95% of nearby stars have "regular" planetary systems similar to our own."[25] Astronomers cannot yet find planets having Earth-like masses using most detection methods, which means such planets also could be quite abundant. (In fact, the first planets ever found orbiting a pulsar do have Earth-like masses.)

One of the more surprising conditions for complex life to evolve listed in table 7.1 is the need to have a Mars-clone as a neighbor. Recall that a Mars-clone is supposedly needed to reseed life on an Earth-like planet after it may have been extinguished during an early period of heavy bombardment from

Figure 7.3 Drawing of a planet discovered around the star 55 Cancri with a hypothetical moon, courtesy of NASA.

210 meteorites and comets. But this condition strangely presupposes that primitive life for some reason *only* could evolve during that early period of heavy bombardment. Although this assumption is without any clear basis, it is reminiscent of the idea that seeds from certain hardwood trees (lodgepole pines) germinate only after devastating forest fires.

Taken collectively most of Ward and Brownlee's conditions for the evolution and survival of complex life seem to be highly speculative and based on pessimistic assumptions. A different set of assumptions might give a much rosier view of the relative chances of finding planetary systems capable of sustaining complex life. The authors of the rare Earth hypothesis are well aware of the speculative nature of their undertaking, as they note: "Lacking knowledge of any extraterrestrial life forms, we cannot be confident we understand the optimal or even the minimal conditions to support life beyond this planet."[26] Yet, despite this admission of a lack of certitude, most of their rare Earth conditions make sense only under a pessimistic interpretation of the evidence.

One astronomer, David Darling, who takes issue with the rare Earth hypothesis believes that Ward and Brownlee may have been overly influenced by Guillermo Gomez, a scientist who has pioneered many of the ideas underlying the hypothesis.[27] In his book *Life Everywhere: The Maverick Science of Astrobiology*, Darling notes that Gomez holds strong creationist beliefs that may have influenced his science. Darling also notes that Ward and Brownlee were apparently unaware of Gomez's creationist writings in developing their hypothesis. The admission by Ward and Brownlee in the preface to their book that "Guillermo Gonzalez changed many of our views about planets and habitable zones" is, therefore, worth noting, but that remark should not be used to discredit the scientific arguments advanced.

The idea that complex life is very rare in the universe has been pushed much further by other authors who have maintained that even one example of a world with complex life is extremely unlikely and, by inference, required divine interven-

211 tion. For example, Hugh Ross as an illustration of the fine tuning necessary for complex life on a planet suggests probabilities for each of 55 independent parameters falling in some "life-friendly" range.[28]

These parameters relate to the kinds of criteria used by Ward and Brownlee in table 7.1.

According to Ross, the joint probability of all 55 parameters being in their allowed ranges is just the product of the separate probabilities, which he quotes as 10^{-69}. Given Ross's further estimate of a total of 10^{22} planets in the entire universe, we find the chances of one Earth-like planet in the universe being only one in 10^{44}—one in a hundred billion trillion trillion trillion. This striking estimate by Ross makes Ward and Brownlee's claims seem positively conservative by comparison. It is difficult, however, to take any such calculations very seriously. As Steven Weinberg has pointed out, the "constants" of nature could conceivably vary in different regions of the universe, and it is only in the region in which we happen to be located that they take on values that make complex life possible.[29] (Also, see the discussion of the anthropic principle in chapter 3.)

In fairness to Ward and Brownlee, we should note that they do mention many of the counterarguments to their REH thesis, even while they come down too firmly on one side. Nevertheless, one cannot help but wonder whether—shades of Goldilocks—they have read too much into the particular circumstances under which life actually evolved on Earth. Surely, if complex life were somehow to evolve under conditions vastly different from Earth, the scientists among them would also conclude that they lived on a world that was "just right."

Are Catastrophes Bad for the Development of Complex Life?

Some of the conditions for the development of complex life summarized in table 7.1 are based on the assumption that a long period free of catastrophes is necessary for the evolution and survival of complex life on a planet. This assumption in turn rests on two others: that complex life is more fragile than primitive life, and that complex life takes a long time to evolve from primitive life. In fact, neither of these assumptions may be correct.

Let's start with the idea that complex life takes longer to evolve than primitive life. Certainly, on Earth complex animal life took billions of years to evolve, but that may not have been the case on Mars. The key issue for the evolution of animal life may simply be a question of how soon an oxygen atmosphere is created. Moreover, we have no idea as to whether primitive life evolves quickly. It could be that Earthly primitive life was seeded by primitive life from space that took billions of years to evolve from nonliving materials elsewhere, and that this first step was the longer process of the two.

The idea that complex life is more fragile than primitive life seems indisputable at the level of individual organisms, particularly when we compare extremophiles with any animal on Earth. But we should not conclude that complex life is necessarily more fragile than primitive life everywhere. For example, on Earth animals have a much narrower range of temperatures under which they can survive, because that's the way evolution happened to occur here, but maybe it didn't have to be that way.

Let's assume that on Earth primitive single-celled organisms evolved near extremely hot undersea thermal vents. There would be little need to be able to tolerate very high temperatures if a multicellular creature evolved the capability to wander away from the vents in search of food. The inability of earthly complex life to withstand very high temperatures may be a simple matter of not having to face that particular environmental challenge. In general, simple organisms are generalists and can survive in a wide range of environments, while complex ones are specialists and thrive in niche environments. Adapting to the environment means less general survival ability when things change drastically.

However, even just limiting ourselves to life on Earth as we know it, it's not clear that complex life *collectively* is more fragile

213   than primitive life. Complex life manages to survive catastrophes by developing enough different species adapting to a wide range of environmental conditions so that, come what may, some of them will survive, even during the worst of the mass extinctions that have occurred.

Many of Ward and Brownlee's conditions for the evolution and survival of complex life were based on the idea that a long period of climate stability is needed for this to occur. For example, the need for a large moon to keep a planet's tilt constant was based on the need to avoid large climate fluctuations. But mass extinctions, though bad for any individual organisms and some species living at the time, may not be bad for the development of complex life as a whole. After each past mass extinction, including the one that killed off the dinosaurs, the number of species, after suffering an abrupt drop, later returned to higher values than before. (A similar effect can be observed with much smaller disasters, such as major wars, economic depressions, and forest fires. Such disasters have beneficial effects, in part because the survivors are stronger than those who perish, and in part because some inefficient or obsolete institutions or organisms are swept away.)

The idea that mass extinctions are actually beneficial for the development of complex life would also seem to be supported by considering the possible vital role played by the two snowball Earth episodes discussed earlier. Thus, paradoxically, complex life might have developed even *faster* on Earth had there been more catastrophic events to serve as the evolutionary stimulus to greater experimentation.

Of course, for this assertion to be true, the mass extinction events need to stop short of those that would destroy 100 percent of all complex life. David Raup has empirically derived a so-called "kill curve" based on how often catastrophes of increasing level of severity have occurred on Earth.[30] For example, he finds that 5–10 percent of all species become extinct every one million years or so, while roughly 100 million years elapse between catastrophes that extinguish 70 percent of all species. Extrapolating Raup's kill curve to the limit, one might

214 conclude that we are living on borrowed time, since 100 percent of all species might be expected to be extinguished after an interval of 2 billion years. However, we cannot know whether such a kill curve is a universal one. It seems plausible that life might evolve hardier forms—leading to a less steep kill curve—on a planet with a harsher climate or more frequent minor catastrophic events.

One way that life might evolve on a planet so as to cope better with even the severest mass extinctions would be if the daily and annual climate variations were considerably greater than on Earth. For example, if the tilt of the planet's axis were greater, and if the eccentricity of its orbit were greater than Earth's, animal life evolving on such a planet would need to tolerate a wider range of temperature variation. Possibly even the cells themselves evolving on such a planet would be of a type that allowed a greater range of temperatures than on Earth. We can only speculate about what kinds of adaptations might occur, but on such a planet, the climate catastrophe of the Earth's axis tipping over on its side might not be a catastrophe at all.

Ward and Brownlee may be right about the snowball Earth episodes being an important stimulus for the development of

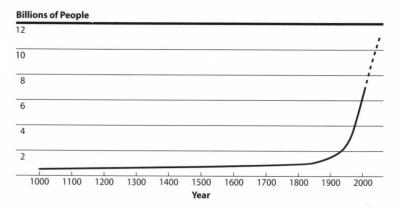

Figure 7.4 Global population. Approximate growth in human population since A.D. 1000.

215 Table 7.2

Creation through Destruction?

Event	Millions of Years Ago	Possible Catastrophic Cause
Origin of life	4000	Heavy bombardment period
Origin of eukaryotic cell	2000	Snowball Earth I
Origin of animals	600	Snowball Earth II
Proliferation of mammals	65	KT asteroid
Origin of artificial intelligence	0	Us!!

complex life, and also their idea that primitive life might have evolved *only* because of the catastrophic early period of heavy bombardment of Earth. In that case, we may conjecture that all the milestones on the road to complex life are the result of catastrophes. However, let's add two more milestones to the list. The first is the flourishing of mammals, which occurred only after the so-called K-T mass extinction that caused the dinosaurs to disappear. Finally, we may add the mass extinction that may be occurring right now, the one associated with the runaway growth in the human population (see figure 7.4), and the consequent destruction of many other species. As usual, organisms alive during a mass extinction see little positive value associated with it. But will this present mass extinction be a positive step forward when viewed by intelligent earthlings a million years from now? If the record of past mass extinctions is any guide, the answer could be yes—even if the intelligent beings doing the assessment are not humans, but our species' descendents (table 7.2).

How Can the Rare Earth Hypothesis Be Tested?

A hypothesis must be capable of being tested to qualify as being scientific. One problem with testing the rare Earth hypothesis is that no quantitative definition is ever given of what "rare" means. Obviously, we would consider Earth to be very rare if it is

216 the only planet in the galaxy capable of sustaining complex life, but suppose it were one of 100, 1000, or 10,000? When would it stop being rare? Testing the first half of the REH, of course, is easy. If primitive life is widespread throughout the universe, we ought to find either present life or the fossils of ancient life throughout the universe, possibly even in our Solar System.

One promising abode for fossils of ancient life is Mars. Several meteorites found in Antarctica have been shown to have a Martian origin, based on the isotopic composition of noble gases found in them agreeing with what is in the Martian atmosphere. A hotly contested claim has been made that one of these meteorites contains fossils of ancient primitive Martian life.[31] Whether the tiny structures found in the meteorite are actually the remains of an organism remains to be proven (see figure 7.5). Certainly, a high priority of any future exploration of

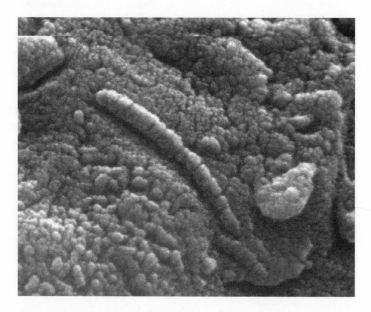

Figure 7.5 Highly magnified image of a putative fossil of an ancient simple Martian organism found in a meteorite identified as ALH84001 that fell on Antarctica. Courtesy of NASA. The width of the photo corresponds to a distance of 1600 microns.

217 Mars will be to determine whether life may have at one time existed there. Further out in the Solar System, looking for life in the subsurface sea believed to exist on the moon Europa could test both parts of the rare Earth hypothesis. If the REH is correct, we might expect to find very primitive life on Europa, but no "complex" worm-like or fish-like critters. Conversely, REH would be disproven if life of any complexity were found there.

Searching for life beyond the Solar System is trickier than searching for extrasolar planets. The over 70 planets found to date around other stars have been found mostly by searching for a small periodic Doppler shift in the spectrum of a star caused by its wobble as one or more planets orbit it. More sensitive detection methods capable of finding the presence of Earth-size extrasolar planets may be possible in the future.[32]

An important method for searching for life on extrasolar planets is to look for planets having an atmosphere that is out of chemical equilibrium, particularly one containing oxygen and another gas with which it would be expected to react (such as methane). The idea is that such nonequilibrium mixtures might imply a continuous production of methane that could signal a living source. This search method will not be implemented for some years, owing to the difficulty of seeing light reflected from a planet in the presence of the enormous background glare from its parent star. In any case, this method of life detection is suited only to finding evidence of primitive life.

Search for Extraterrestrial Intelligence (SETI)

When most people think of complex life elsewhere in the universe, the main question on their minds is aliens having a sufficient degree of intelligence that we might have something in common. The discovery of "complex" aliens at the level of sponges would not be nearly as interesting as the discovery of intelligent life elsewhere (figure 7.6). The ongoing SETI project—now funded privately after federal funding was cut off—is an effort to search for extraterrestrial intelligent life

218 based on signals that aliens might be broadcasting. Even though interstellar distances would probably make any back-and-forth "conversation" unlikely, the mere detection of signals from an alien civilization would be one of the greatest scientific discoveries of all time. Thus, SETI is a project worth pursuing, even if the chances of success are completely unknown.

One author has claimed that the search for a SETI signal is an example an unfalsifiable hypothesis, which can never be proven wrong, and therefore SETI is not worth pursuing.[33] In

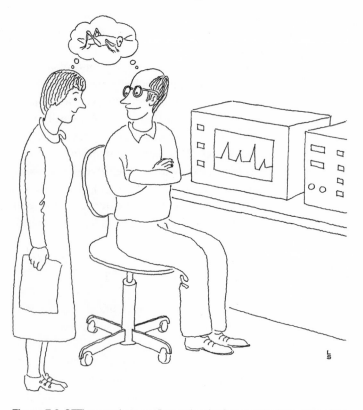

Figure 7.6 SETI researchers on discovering the first steady chirp–chirp–chirp wonder about the identity of the transmitting civilization.

219 other words, if no SETI signal is found, we can never *dis*prove that there are intelligent aliens, because any number of reasons exist, as noted below, why we might not get a signal, so the hypothesis of intelligent aliens out there is not testable. This criticism of SETI research is unfair. It is true that the absence of a signal doesn't prove nobody's out there, but researchers might actually find a signal. And *that* is the whole point of the search. SETI may be easy to ridicule, but it is not outside the bounds of real science. Clearly, the amount of money best spent on SETI will depend on some combination of our estimate of the probability of success (unknown, but probably low) and the impact of finding a SETI signal (very significant).

Suppose after years of searching for signals, none are found—how much support would that give to the rare Earth hypothesis? Surprisingly, the answer may be not very much. We can imagine any number of reasons why the lack of detection of a SETI signal need not imply that extraterrestrial intelligent civilizations are very rare. Here are some possible reasons.

The SETI signal is there, but it is . . .

- too weak to be picked up by our primitive receivers.
- transmitted at a frequency outside the range we are scanning.
- transmitted as a bit-stream without a single frequency carrier.
- transmitted using hypothetical faster-than-light "tachyons."
- emitted by planets of stars we haven't yet pointed receivers at.
- only beamed toward worlds that aliens think are worth their while. (Maybe, they've received our TV broadcasts!)

Alternatively, the aliens don't send out signals because they've . . .

- concluded that broadcasting their location is unwise.
- given up transmitting after some years trying.
- no interest in conversations with centuries between transmissions.

• insufficient resources to invest in such very costly projects.
   • no technology, even though they are intelligent.
   • evolved on a cloud-covered world and have never seen the sky.
   • not developed any curiosity about life elsewhere.

You can come up with some additional reasons, I'm sure. Some of the preceding reasons are more serious than the others, but none of them can be entirely ruled out—even the idea of transmitting signals using hypothetical faster-than-light "tachyons."[34] Thus, the nondetection of a SETI signal may not say a great deal about whether intelligent life is out there. (For example, if by chance these preceding conditions should eliminate 99.9 percent of intelligent civilizations in our galaxy, we might not get any signals even if the galaxy had a thousand civilizations.)

Moreover, even if a SETI signal were eventually found, we still might not know if the REH is correct. As fundamental as the discovery that we are not alone would be, the reception of one signal would tell us only that the number of intelligent technological worlds in the universe was greater than one. (On the other hand, it might be argued that only a small fraction of complex life evolves to the point of intelligent life capable of transmitting signals. In that case, we might infer that for each extraterrestrial communicating world we found, there might be thousands of others with complex life in the form of lizards or sponges, which would imply that complex life is quite abundant.)

A more interesting speculation concerns the possibility of extremely advanced aliens. More than likely, any alien civilizations out there capable of sending us messages would be so far ahead of us that their bothering to try to communicate with us would be as futile as our trying to communicate with lizards—OK, maybe dogs. This speculation depends greatly on how long intelligent civilizations exist before destroying themselves. If the average lifetime were appreciably more than a thousand years, then it seems unlikely any ETs out there would

221 be as close as we are to the dawn of the technological era, when it first became possible to transmit radio signals. To avoid their own destruction, it's quite possible that the ETs are particularly advanced in their interpersonal intelligence (see chapter 4), rather than their logical/mathematical ability. Who knows? Maybe they have even evolved to the point where they possess abilities that we would consider paranormal—see chapter 5.

Finally, let's return to the rare Earth hypothesis, which claims that complex life is very rare in the universe, even though simple life is likely to be abundant. Given the highly speculative nature of the evidence for the rare Earth hypothesis and its lack of specificity in terms of just how rare complex life is in the universe, I'm inclined to rate the idea as 2 flakes.

8 Can a Sugar Pill Cure You?

He cures most in whom most are confident.

—*Claudius Galenus (Galen) of Rome (131–200 A.D.), whose drugs and remedies were used extensively for fifteen hundred years*

New medicines should be used as quickly as possible before they lose their power to heal.

—*a recommendation of nineteenth-century physician Armand Trousseau*

NOTHING ILLUSTRATES the intimate connections between mind and body better than the mysterious placebo effect. One dictionary defines placebo as "a medication prescribed more for the mental relief of the patient than for its actual effect on a disorder." Here we will take the placebo effect to mean the improvement that results in your condition following some treatment that objectively should have no effect, such as an inert pill. As indicated in the first of the two quotes opening the chapter, treatments—either real or placebo—seem to work best when both the physician and the patient have confidence in them (*and* in the physician).

There are as many ways to interpret the second quote as there are reasons medicines might lose their effectiveness over time, including the loss of potency of active ingredients and the development of new strains of drug-resistant bacteria. But we could also interpret the second quote to mean that many new drugs were in fact placebos until the age of scientific medicine,

223 and their effectiveness waned when physicians lost confidence
in them. In other words, in many cases, what was initially considered to be a "real" drug later turned out to be nothing more
than a placebo. One might imagine that in the age of scientific
medicine such cases no longer occur, but, as we shall see, that
may not be true.

Even bona fide drugs, such as antibiotics, can be considered
placebos if, for example, they are given to patients in situations
for which they are known to be ineffective, such as remedies for
colds, which are caused by viruses not bacteria. Placebos, of
course, are not limited to drugs, but include any sham treatment, such as sitting a patient in front of a therapeutic machine
when it is turned off or using fake magnets instead of real ones
in "magnet therapy." This last example begs the question of
whether certain "alternative" medicine treatments such as
magnet therapy are themselves, in fact, simply placebos.

The term placebo entered the language in the nineteenth century, based on a translation of a Latin verse from vespers, a
chant for the dead, meaning "I will please." (The connotation of
the phrase is "I will please *by deception*.") Placebos, and the idea
of physicians prescribing medicines to keep their patients
happy, have a long history in medicine. In fact, it has been suggested by Shapiro and Shapiro, two physicians who have studied placebos extensively, that "the history of medical treatment
until recently is largely the history of the placebo effect."[1] This
controversial hypothesis flies in the face of some opinions that
many useful remedies can be found in the folklore medicine
compiled over the ages. However, according to the Shapiros,
remedies passed down from the ancient world include nearly
5000 drugs, and "with only a few possible but unlikely speculative exceptions, all were placebos."

In fact, many of the remedies before the age of scientific medicine were worse than placebos and were definitely harmful,
such as blood letting, accupuncture using unsterilized needles,
poisonous drugs, and unnecessary surgery. Medicine has made
giant strides forward in the last 150 years, following the discovery that germs cause many diseases and subsequent advances

224 in biochemistry and genetics, but most remedies continued to be placebos until well into the twentieth century. In fact, even as late as 1950 any given medicine was used as a placebo an estimated 40 percent of the time, according to an editorial in the *British Medical Journal*.[2]

   At one time many physicians routinely kept a supply of inert placebo pills of all sizes shapes and colors to give to patients with vague symptoms and for whom they couldn't diagnose any organic illness. Other physicians have regarded that practice as repugnant, perhaps because it served to remind them of medicine's prescientific roots, and the many mysteries that still underlie the healing process. Most physicians believe that the ability to distinguish "real" from placebo treatments is the key to scientific medicine. This ability did not exist reliably until medicine adopted the so-called *double-blind* method of testing new drugs or procedures, in which placebos play a crucial role as the medication given to a control group. In some studies an older proven drug is used as the medication given to the control group instead of a placebo, and in other studies there are three comparison groups: new drug, old drug, and placebo. Generally, for ethical reasons, no treatment group in a study may be placed at a disadvantage—a principle known as clinical equipoise. Therefore, a placebo group is generally acceptable only if there is no available treatment for an illness or if the treatment could be withheld without harm to the patient. (In practice, trials are often conducted that partly violate these principles.)

The Double-Blind Method

The double-blind method was first introduced in a study by W. H. Rivers in 1908 to look at the effects of alcohol on fatigue.[3] The idea of the double-blind method is that when a new drug is compared with an existing one (or a placebo), neither the physician nor the patients know who is receiving the new and old drugs. That way the beliefs of doctor and patient about the efficacy of the two drugs cannot influence the outcome of the study. A study is considered single-blind if only one of the parties (usually the physician) knows which group the patients are in.

The double-blind method was strongly resisted for many years, perhaps because it showed that some favored and widely accepted treatments were, in fact, no more effective than placebos. One classic case involved a surgical procedure known as ligation of the mammary artery used to treat angina pectoris. Angina is a recurrent chest pain due to reduced blood flow to the heart. The ligation surgery was reported to be 68–90 percent effective in two *un*controlled studies. However, when it was compared to "placebo surgery" involving small incisions, in a blind study, the procedure was found to be less effective than placebo and was abandoned.[4]

Many other surgical procedures, including tonsillectomy and hysterectomy for treatment of "hysteria," have also been shown to have no benefit or to be no more effective than placebos. In one recent double-blind trial involving 300 patients a laser surgical procedure to unblock clogged arteries was shown to be no more effective than placebo surgery.[5]

It is only speculation to suggest reasons why physicians took so long to adopt the double-blind test as the gold standard of scientific medicine, but surely their reluctance to face their own fallibility was a factor. It was not pleasant for physicians to admit that they often saw what they wanted to see and heard what their patients thought they wanted to hear. A large factor may also have been physicians' underestimation of the amount of spontaneous improvement that can happen without them.

Such was the resistance to accepting the double-blind method as a standard that it was not until end of the 1970s that the U.S. Federal Drug Administration required the use of the double-blind method for approval of most new drugs.[6] Faithful mimicry of the real treatment by the placebo is essential for the double-blind method to be reliable, since, otherwise, patients and doctors may know or guess which group they are in. The study then becomes "unblinded" and possibly biased.

The Powerful Placebo

Estimates of the effectiveness of placebos vary widely according to the condition they are used to treat. As might be ex-

226 pected, placebos work better on imaginary (or imagined) ill-nesses than on diseases that have an organic basis. But that does not imply that when placebos cure an illness there was no physical basis for the illness or the cure.

. The effectiveness of placebos also depends on the nature of the placebo, the patient, and many other factors. For example, the type of situations where placebos are found to be most ef-fective involve

- *doctors* who are sympathetic, enthusiastic, and caring, and believe in the treatment.
- *patients* who are optimistic, believe in the doctor and treat-ment, and are capable of adhering to a treatment regimen.
- *placebos* that are credible and suited to the patient and the symptoms.
- *symptoms* that are minor, subjective, and vary over time, such as depression, anxiety, and, especially, chronic pain.

The importance of the psychological dimension in the expe-rience of chronic pain is widely known in the medical field. One old joke involves a physician asking a patient with chronic pain: "What have you done that you should be pun-ished in this way?" While the size of the placebo effect de-pends on some characteristics of the patient, it does not appear to depend on others. Studies show that it exhibits no depend-ence, for example, on intelligence or personality type.[7] Rather, much more of a placebo's effectiveness seems to depend on the setting in which placebos are used and one's individual and cultural milieu.

Thus, you might react either favorably or unfavorably to a placebo depending on the setting, your relationship with your doctor, the meaning of the placebo in your culture, etc. The im-portance of the cultural context is stressed by medical anthro-pologists, who believe that placebos work partly because they are symbols of powerful healing rituals. Although most Western physicians might be loathe to admit that they have anything in common with healers in primitive societies, the medical anthropologists believe otherwise. These researchers

227   assert that the healing rituals in other cultures are sometimes
quite effective, even though they are mostly based on the
placebo effect.

Medical Anthropology

Medical anthropology is the study of the cultural influence on medicine
and disease. According to medical anthropologists, even the very definition of what constitutes a disease is culturally based. In Donald
Joralemon's *Exploring Medical Anthropology* two examples nicely make
this point.[8] In eighteenth-century Europe and the United States physicians were concerned about a disease known as "onanism," which was
considered to be responsible for epilepsy, blindness, vertigo, loss of
hearing, memory loss, and assorted other ailments. Today we know
onanism by the term masturbation! Joralemon cites an equally bizarre example from nineteenth-century America—the disease known as drapetomania. This ailment, and the term for it coined by Samuel Cartwright, a
southern physician, is the disease of mind that "induces the negro to run
away from service." Dr. Cartwright assured readers that with proper
treatment drapetomania can be cured, even if slaves were located near
the borders of free states. How certain can we be that some of our present day ailments, such as multiple-personality disorder and andropause
(male menopause), are not as fictitious as these bizarre examples?

The title of this section, "The Powerful Placebo," is that of an
article by Beecher that contained the first estimate of placebo
effectiveness as 35 percent, meaning that on average this percentage of patients report relief of symptoms when receiving a
dummy treatment.[9] A more recent meta-analysis of 55 studies
estimated that placebo reactions were in the range 24 to 58 percent, with the highest effect for the treatment of pain.[10] Drugs
used to treat depression also show very high placebo effects.

In some experiments the size of the placebo effect has been

228 large enough even to overcome an active drug and actually reverse the direction of the outcome. In one experiment done in 1950 a pregnant woman who complained of nausea was given a drug that her physician said would control it. Unknown to the woman the substance was ipecac, which normally has the effect of inducing vomiting, but in this case it had the opposite effect.[11] (This experiment would not be repeated today both for ethical reasons and its questionable methodology.)

The Powerless Placebo?

Given the increasing public awareness of the significance of the placebo effect, it was surprising when a pair of Danish researchers reported in a 2001 study that the placebo effect was vastly overblown. In the words of one skeptical reviewer of the study, "To be told today that placebos are powerless after all is something like being told that the human genome is a myth just after scientists had announced that they had finally been successful in mapping it in its entirety."[12]

But the two Danish physicians who did the study, Asbjorn Hrobjartsson and Peter Gotzsche (whom we shall refer to as H&G), had observed an important flaw in most previous analyses on the effectiveness of placebos.[13] Since the use of placebos in most drug trials is merely that of a control to test some active drug, researchers often have an additional "no treatment" group. No one previous to H&G had done a meta-analysis that compared the group who received placebos against patients receiving no treatment, which is the appropriate control group if the placebo is the "treatment" whose effectiveness is being studied.

This H&G observation of the flaw of earlier placebo studies is extremely important, because a substantial fraction of the placebo effect might simply relate to the natural course of illnesses whose symptoms fluctuate over time. A patient who has a back pain of varying severity would be most likely to visit a doctor when the pain is at its worst. As a result, any treatment

229 the doctor gives, even a sham placebo treatment (*or no treatment at all*) is likely to be followed by a lessening of the pain, simply because of natural fluctuations of the pain about its average or mean value. This tendency, known as "regression to the mean," occurs in many areas of life, from the stock market to the inheritance of genetically determined characteristics. It was first pointed out by the statistician Francis Galton, who noted that children born to above average height parents are, apart from the effects of nutrition, more likely to be shorter than their parents, i.e., closer to the mean height.[14] (That is not to say that tall parents do not tend to have taller than average children—only that their children tend to be intermediate in height between that of their parents and the mean of the population.)

The tendency of many symptoms to wax and wane about some mean value is probably also responsible for the apparent efficacy of many alternative medicine fads. Suppose, for example, that when your pain is at its worst, you put on a copper bracelet. Most likely your pain would then spontaneously subside back to some average value, and it would be easy to assume that the copper bracelet had something to do with the decrease in pain. Your continued belief in the bracelet's power may even be able to keep the pain at bay in the future, since it had already "demonstrated" its therapeutic power. As this example illustrates, the placebo effect probably consists in part of a simple regression to the mean (due to natural fluctuations in symptoms), but also could have a "more interesting" component based on psychological factors. The real questions are how large are these two respective components and how can they be separated from one another empirically?

H&G were able to resolve this issue in their meta-analysis of prior studies that compared the effect of placebos to that of "no treatment." Essentially, they did a search for all published studies in which these two terms ("placebo" and "no treatment") appeared, and imposed several other conditions, such as studies having an acceptably low dropout rate. The definition they used for "placebo" was whatever the authors of each study intended it to mean; most commonly it was an inert pill. H&G's

230 search of the relevant databases yielded 114 separate studies, most of which included placebo and no treatment groups merely as controls used to evaluate some active drug or treatment.

The most common problem treated was pain (involved in 27 studies), although a total of 40 different conditions, including obesity, asthma, and high blood pressure, were examined. It is unclear if it makes a great deal of sense to examine the placebo effect in the aggregate for such a diverse range of conditions, since the size of the effect might be expected to vary depending on the condition treated. But the number of studies for any one condition was often too small to say anything definitive. H&G therefore found it useful to combine studies involving different conditions to gain statistical power.

One type of combination was to aggregate all studies having only two possible (binary) outcomes. Typical binary outcomes might involve whether patients either did or did not stop smoking, did or did not lose weight, or perhaps whether they were alive or dead(!) at the end of a trial. Less than a third of the 114 trials involved binary outcomes, and when these were lumped together the result was that no distinction was observed between the outcome for the placebo groups and the untreated groups. Thus, H&G concluded that for binary outcomes the whole of the placebo effect is due to regression to the mean, or the natural variation of symptoms over time! The statistical power of their analysis was such that if a placebo effect were present in these pooled binary trials (apart from simple regression to the mean), it would have to be less than 12 percent (with 95 percent confidence).

H&G also examined the combination of all trials involving continuous rather than binary outcomes. For many conditions, such as obesity or high blood pressure, continuous outcomes are more relevant than binary ones to the health of individual patients. For example, physicians arbitrarily could measure a weight-loss treatment's effectiveness by the binary outcome of whether or not it reduced each patient's weight by at least 10 pounds. But it seems unlikely that individual patients whose

231 weight was reduced by 9.9 pounds in the trial would regard their treatment as significantly less successful than those who lost 10 pounds.

To measure the size of the placebo effect for continuous outcomes, H&G defined a "standardized mean difference" (DIFF) between the placebo and no treatment groups. The definition is such that DIFF = −1.0 would mean that the placebo groups would be one standard deviation better off than the no treatment groups. The DIFF they actually found for the aggregate of all continuous outcomes was −0.28 (with a 95 percent chance of being in the range from −0.38 to −0.19). What that result means is that the placebo groups involved in trials with continuous outcomes had between 19 and 38 percent more of a benefit than the no treatment group.

As seen in table 8.1 from H&G, with the exception of pain (whose 95 percent confidence interval is entirely below zero), no other specific condition was found to show a statistically significant reduction in the continuous outcome category. Only conditions present in three or more trials are tabulated.

As the table shows, there was no statistically significant result for the five individual conditions tabulated besides pain, because the 95 percent confidence interval (CI in the table) straddles zero in those cases. However, in all five cases the negative sign of DIFF shows that the direction of the result is the same: placebo groups have less of the unwanted condition than the untreated groups. It therefore seems likely that were it not

Table 8.1

Effect of Placebo on Specific Clinical Problems

Outcome	Participants	Trials	DIFF	95% CI
Pain	1602	27	−0.27	−0.40 to −0.15
Obesity	128	5	−0.40	−0.92 to 0.12
Asthma	81	3	−0.34	−0.83 to 0.14
Hypertension	129	7	−0.32	−0.78 to 0.13
Insomnia	100	5	−0.26	−0.66 to 0.13
Anxiety	257	6	−0.06	−0.31 to 0.18

232   for the limited number of trials for all but the condition of pain, a more statistically powerful test might indeed show an effect for these other five conditions. Also, note that two of these five conditions (obesity and hypertension) are objectively measurable by an outside observer. Overall, when all objectively observable conditions are lumped together the result is not statistically significant, however. *This led H&G to their overall conclusion that placebos have not demonstrated to be effective except for conditions having a subjective continuous outcome, most especially pain.*

Many criticisms have been leveled at the H&G study, not all of which are justified. Clearly, the limited number of trials for most of the 40 different conditions makes it difficult to see an effect in those conditions individually. As already noted, the data in table 8.1 are suggestive of an effect in the five conditions other than pain had more trials been available. Another problem with the H&G study (which they acknowledge) is that their study uses a "limited" definition of placebo. They did not examine the effect of the patient–physician relationship, and they "could not rule out a psychological therapeutic effect of this relationship." That study limitation is quite important because even the untreated group in a clinical study is interviewed by a physician periodically and could be said to benefit from a type of placebo effect merely as a result of being enrolled in the study. Finally, it is unclear from the H&G study how many in the "no treatment" group really received no treatment. The most severely afflicted members of this group might have sought treatment elsewhere or might have dropped out of the study. Both of these possibilities would make the average of the remaining members of the no treatment group better off relative to the placebo group, and result in a lower placebo effect.

In summary, the H&G study may have shown that the placebo is not as powerful as some observers would believe, but it certainly is far from powerless. The limitations of their study do not justify a conclusion that the almost all of the placebo effect (understood in the broader sense of including the patient–physician interaction) can be attributed to simple re-

233 gression to the mean. In fact, it could be argued that the H&G
analysis actually says little about the placebo effect, since that
term does not appear in their paper. As noted earlier, H&G
merely compared the extent of symptom reduction in placebo
and active treatment groups, with the former defined however
the original study authors chose to define it.

What should we make of H&G's final suggestion in their paper that "the use of placebo outside the aegis of a controlled,
properly designed clinical trial cannot be recommended"?
Clearly, a physician's use of a placebo to treat a patient does
raise a host of ethical and legal questions. Such usage would be
inappropriate if effective nonplacebo remedies existed. But in
cases where the conventional remedies are painful or costly,
have adverse side effects, and are little more effective than
placebos, perhaps placebos could be justified, particularly if the
condition is not a serious one. There are clearly some therapeutic examples where placebos might be useful, especially for
subjective conditions such as chronic pain, where H&G themselves have recognized a benefit. One condition among many
less serious ailments where placebos might be useful is that of
warts.

The Interesting Case of Warts

Placebos have been shown to be quite effective in treating skin
warts, which are clearly not a subjective ailment and are caused
by viruses. According to an Australian physician, F. E. Anderson,
warts probably have the highest number of folk remedies of any
disease, which is not surprising if they respond well to placebos.[15] Some of the stranger folk remedies include "selling" them
to someone who carries them away, rubbing them with pork fat
(suggested by Francis *Bacon*!), or secretly rubbing them on the father of an illegitimate child.[16] Warts are more common in children than adults, with the peak occurring between 12 and 16
years of age. They can occur anywhere on the body, but most
commonly appear on the hands.

234   Two-thirds of skin warts disappear spontaneously in about two years, for reasons that are not understood. They are often treated using a variety of topical agents applied to the skin, and also by surgery. One technique for eliminating warts that has been used with some success is hypnotic suggestion, which some might consider a placebo treatment. Some critics of the hypnosis method suggest that we cannot rule out the possibility of spontaneous recovery following the hypnosis, but the same could be said about any treatment method, and the cure rate for hypnosis reported in some studies is comparable to that with surgery. One classic study involved patients having warts on both sides of their body. The British researcher A.H.C. Sinclair-Gieben used hypnotic suggestion to treat the warts on one side of the body—the side on which the warts were most severe. After 5 to 15 weeks of hypnotherapy the warts on the treated side of the body disappeared while the warts on the other side remained in 9 out of 10 patients.[17] This result is very difficult to reconcile with the possibility of a spontaneous recovery. In one interesting and effective placebo treatment of warts inert colored water is used to coat them. Of course, the patient is not told that the treatment is a placebo, and the treatment should be administered in a believable manner.

A personal digression here will explain my particular interest in the use of placebos to treat warts. Over a half-century ago I was a shy, awkward teenager whose hands were completely covered by ugly warts. The condition, which lasted for several years, was acutely embarrassing to me, often leading me to keep my hands in my pockets. My favorite aunt, Gertie, apparently had learned somehow that ordinary colored water was able to cure warts. Gertie lovingly coated my warts every few days with this "medicine" that she kept in a labeled bottle. After two weeks my warts miraculously all disappeared.

I do not consider myself a particularly suggestible person, either now or then. It is possible that the disappearance of the warts was spontaneous and not the result of Aunt Gertie's "medicine," but in that case the timing would have been very coincidental. This experience made a deep impression on me, especially after I learned that the medicine was just colored

235 water. Thinking back on the matter now it seems plausible to me that not only the placebo treatment, but perhaps the condition of warts itself has an important psychological dimension. The question "what have you done to be punished in this way?" has great resonance to a shy teenager awakening to the mysteries of sex and his own body, who would have had ample reason to make his hands look ugly because of his guilt. It may not be an accident that skin warts most commonly appear on the hands, and that the peak age of occurrence is 12 to 16 years.

Placebos in the Mental Health Field

Although many studies have shown that psychotherapy is more effective than a placebo, it is very difficult to conduct a study with adequate placebo controls in this field. For example, a neutral monthly discussion could hardly serve as a placebo for a weekly hour-long psychotherapy session. As a result, perhaps the best one can do is to compare one type of therapy with another. Unfortunately, none of the seven studies of this type show that any one form of therapy (from watching Marx brothers movies to Freudian therapy) is better than any other or better than a placebo.[18]

Even more surprising, the therapeutic value of various forms of therapy (now numbering over 250) has not been shown to depend on any relevant variables one might expect, including the length of session, the number of sessions, the type of problem treated, or even the training or experience of the therapist.[19] Amiable college faculty produce about the same results as expert psychotherapists. It's enough to make one wonder if all of psychotherapy is one big placebo? To quote the slightly more diplomatic Shapiro and Shapiro: "Although there is general agreement that psychotherapy is useful, beneficial, and effective for many patients, as is true of many notable placebo treatments, the knotty question remains—is psychotherapy more than placebo?"[20]

Obviously, these controversial assertions are not unchallenged by other psychologists. For example, a 1995 study, while

236 conceding that no specific type of psychotherapy did any better
than any other for any disorder, concluded that patients nevertheless benefited very substantially from the treatment. It also
found that certain care providers (marriage counselors and
family doctors) were not as effective as trained psychologists
and psychiatrists, and that patients receiving long-term treatment did benefit more than those receiving short-term treatment.[21] And yet the Shapiros' "knotty question" about psychotherapy being no more than a placebo remains unanswered.

Eliza—The "Computer Therapist"

In 1966, Joseph Weizenbaum, an artificial intelligence researcher at MIT,
created one of the most famous AI programs ever written. The program,
known as Eliza, was written to mimic the responses that a therapist might
give to questions typed in by a patient. The responses were such that
many people interacting with the program had the sense they were actually in contact with a real human who was concerned about their problems. In fact, Weizenbaum was surprised (and horrified) to learn that
many people seemed to form strong emotional bonds with his creation
during their sessions and that some psychiatrists were even ready to begin letting Eliza treat their patients! The program actually understands
nothing that is said to it. It works by simply "twisting the statements of her
patients back at them in the classic manner of a nondirective psy-
chotherapist."[22] A version of Eliza, the "computer therapist," is available
free on the internet, and can be found at the web site: www.peccavi.
com/eliza/. A sample exchange between the Eliza program and a live
teenager is given below. Successors to the Eliza program, such as PC
Therapist from the Thinking Software Company, mimic interactive human
conversation even more closely than Eliza. PC Therapist, in fact, was the
first computer program to pass the so-called Turing test by convincing
five out of ten judges that its responses to their queries were those of a
human rather than a computer. I am unaware of any controlled studies
that compared computer-only therapy with conventional treatment.
However, one 2001 review of the literature looked at studies that had
been conducted on the effectiveness of computer-aided therapy. It found

237 that people receiving computer-aided treatments had as good or better outcomes as people in control groups.[23]

Here is a sample dialog with Eliza, whose responses are capitalized, from Copeland[24]:

Men are all alike.
IN WHAT WAY
They're always bugging us about something or other.
CAN YOU THINK OF A SPECIFIC EXAMPLE
Well, my boyfriend made me come here.
YOUR BOYFRIEND MADE YOU COME HERE
He says I'm depressed much of the time.
I AM SORRY TO HEAR THAT YOU ARE DEPRESSED
It's true that I am unhappy.
DO YOU THINK COMING HERE WILL HELP YOU NOT BE UNHAPPY
I need some help, that much seems certain.
WHAT WOULD IT MEAN TO YOU IF YOU GOT SOME HELP
Perhaps I could learn to get along with my mother.
TELL ME MORE ABOUT YOUR FAMILY
My mother takes care of me.
WHO ELSE IN YOUR FAMILY TAKES CARE OF YOU
My father.
YOUR FATHER
You are like my father in some ways.
WHAT RESEMBLANCE DO YOU SEE
You are not very aggressive but I think you don't want me to notice that.
WHAT MAKES YOU THINK I AM NOT VERY AGGRESSIVE
You don't argue with me.
WHY DO YOU THINK I DON'T ARGUE WITH YOU
You are afraid of me.
DOES IT PLEASE YOU TO BELIEVE I AM AFRAID OF YOU
My father is afraid of everybody.
WHAT ELSE COMES TO MIND WHEN YOU THINK OF YOUR FATHER
Bullies.
DOES THAT HAVE ANYTHING TO DO WITH THE FACT THAT YOUR BOYFRIEND MADE YOU COME HERE

238   Even if all of psychotherapy is just a placebo, one type of treatment that we would assume is not would be the active drugs that are used to treat depression and other mental illnesses. As we shall see, that assumption may be in error. (An objective analysis of the many difficulties inherent in evaluating the effectiveness of these drugs is given in Fischer and Greenberg's *The Limits of Biological Treatment for Psychological Distress.*[25]

Economics and Marketing of Antidepressant Drugs

The issue of marketing antidepressant drugs may seem to have no relevance to placebos. But the immense profitability of these (and other) drugs can lead to enormous pressures to approve certain drugs that, in fact, may have little or no advantage compared to placebos. Antidepressants are the most costly of the psychotropic drugs. For example, as of this writing a dose of Prozac costs $2.47, which may be compared to the cost of its active ingredients, which is only 0.11 cents.[26] It might be objected that such a comparison (and the implication of a 225,000 percent price markup) is unfair, because it ignores the costs of research. On the other hand, research costs are less than the costs associated with marketing the drugs. According to a study of drug manufacturer's expenditures, their research and development budget is less than half that spent on marketing and administration and the gap continues to grow rapidly in favor of marketing.[27] Moreover, that marketing is increasingly directed toward the individual consumer. According to *Industry Intelligence:* "It is predicted that top tier pharmaceutical companies will increase their DTC [Direct to Consumer] spend[ing] from 15% of their marketing budget in 1998 to an estimated 30% in 2000 and 45% in 2005."[28] Amazingly, some of the television advertising for new drugs doesn't even say what condition the drugs are intended to treat, and simply suggest: "Ask your doctor if XYZ is right for you."

All that marketing appears to be working. The number of doctor visits for depression, for example, has nearly doubled in the United States since 1987.

239 Many patients visiting doctors for what they think is a physical problem also suffer from depression, which may be the root cause of their physical ailment. It is therefore understandable that physicians have been encouraged to ask patients questions to screen for depression during regular physicals ("Have you felt down, depressed, or helpless during the past two weeks?").

And yet despite the obvious laudable goal of a desire to alleviate suffering and track down root causes, one cannot help but wonder if drug companies have also found one more ingenious avenue for their marketing efforts. In fact, why stop the marketing effort in seeking out depressed humans? Some veterinarians have begun recommending antidepressants for pets that are behaving badly. Perhaps your dog exhibits "X-files" (unexplainable) behavior symptoms? Perhaps your cat attempts to "rape" other cats? In cases such as these author and veterinarian Nicholas Dodman suggests that antidepressants may fix the problem.[29]

Figure 8.1 Drawing provided by Barrie McGuire, printed with permission.

240   The question of interest here is the strength of the placebo effect for psychotropic drugs, especially the antidepressants—the most prescribed of all the psychotropic drugs. In other words, we wish to examine the evidence that these drugs are in fact more effective than placebos. Actually, even if placebos were only 75 percent as effective as the real drug, as one study of 2300 patients claims, you might still prefer receiving an inert pill costing 0.11 cents compared to one that costs $2.47.[30]

Surprisingly, the evidence that antidepressants are superior to placebos is weak. For example, one 1998 meta-analysis showed that in only two out of nine trials examined did the active drug show a larger effect in relieving symptoms than the placebo comparison by a statistically significant amount, i.e., to a 95 percent confidence level.[31] This meta-analysis has been criticized unfairly for relying on older trials and examining only nine of them. However, as we shall see, the small number of trials used and their older age speaks ill not of the researchers who compiled this meta-analysis, but rather of the general practices followed in the field.

The researchers, in fact, used not all the trials available at the time, but, wisely, only those using "active" placebos that mimic drug side effects. The use of active placebos makes it more difficult for study participants to know whether they are in the drug or placebo group and helps ensure that the study remains blinded. Apparently, the small number of trials in the meta-analysis (nine) and their older dates (all but one before 1976) is a reflection of the fact that active placebos have been used very sparingly in recent antidepressant trials.

The problem of unblinding is an important one, because studies show that with passive placebos between 78 and 88 percent of patients and physicians in antidepressant trials can correctly identify whether the drug being administered is the placebo or the active drug, based on the presence or absence of side effects and other subtle cues.[32] Obviously, if you strongly suspect that you are in the placebo (or active drug) group, that belief will negatively (or positively) impact the strength of the placebo effect. This unblinding effect will bias the study by giving an inflated relative performance of the "active" drug.

241   It is not surprising, for example, that when patients in a placebo group were informed at the end of one study that they had been receiving dummy pills their condition quickly deteriorated. Even if blindness is breached only that they had been receiving dummy pills for the physician and not the patient, the study can be become biased. Research has shown that a physician's expectation can affect the outcome of the study, especially in psychiatry, where many judgements are subjective.[33] In an appendix at the end of this chapter a mathematical model is presented that shows exactly how the unblinding phenomenon works. Specifically, we show how it can yield a positive statistically significant result in a clinical trial of a drug that is no more effective than a placebo.

Drug companies face an interesting quandary with regard to drug side effects. Clearly, if the side effects are too serious or annoying they will impair the ability of the drug to receive approval, or the potential market after approval. If the side effects are mild, but easily recognizable by patients, then they can be a major factor in unblinding the study and make the drug appear much more effective than it actually is, because patients in the placebo group will become aware they are taking dummy pills. There is, therefore, great economic value to drug companies in discovering new drugs that have easily identifiable but not serious side effects. Barring a revelation by an industry insider, one can only speculate about what fraction of the research on new drugs involves the search for such "desirable" side effects as dry mouth that would permit the drugs to be easily recognized by participants in clinical trials.

As noted previously, virtually all antidepressant trials nowadays use passive placebos that don't mimic antidepressant side effects. But even with trials involving passive placebos (where unblinding is a major source of bias), the performance of active drugs relative to placebos is not particularly impressive. In an analysis of 96 trials done by Arif Khan, placebos and active drugs were statistically indistinguishable in 52 percent of trials.[34] This is a more impressive performance than the two trials out of nine noted earlier, but the problem of unblinding could easily skew the results in favor of the active drug and

242 against the placebo. Moreover, the correlation between the effects of placebo and the active drug seen in one meta-analysis of 19 trials of antidepressants was 0.9.[35] This high correlation means that the size of the placebo effect mirrored to a very high degree that of the active drug in those studies (figure 8.2). According to the authors of that meta-analysis, the high correlation suggests that the "active" drugs tested in the 19 trials are essentially just placebos. Recent research using PET scan images of the brain suggests that antidepressants and placebos even have similar effects on the brain.[36] Both the active drugs and placebos increased the blood flow in those areas of the brain known to be rich in opioid receptors, as shown in figure 8.3.

Aside from the relief of symptoms in trials of new antidepressant drugs versus placebos, a more serious life-and-death issue concerns their relative effects on rates of suicide and attempted suicide. Table 8.2 shows the data compiled by Arif Khan for both new and established drugs versus placebos from two studies involving a total of 30,000 patients, among whom there

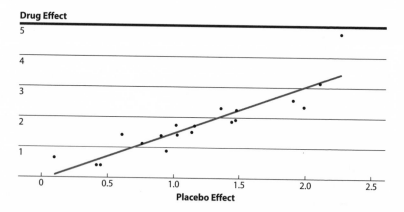

Figure 8.2 Drug versus placebo effect. Relative sizes of drug and placebo effects in nineteen studies of antidepressants. The effect sizes for drug and placebo groups are found from the difference in the mean depression score before and after treatment divided by the standard deviation. Reprinted with permission from Kirsch and Sapirstein.[35] Copyright © 1998 American Psychological Association.

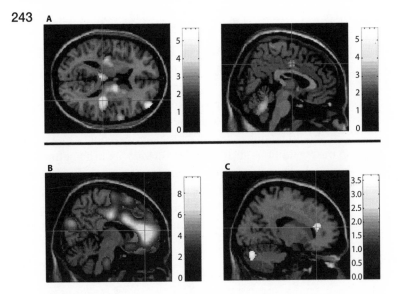

Figure 8.3 PET scans showing brain activity under various situations. Increased brain activity is shown by light areas in the same brain regions when subject is taking active antidepressant medicine (B) or placebo (C). The two upper scans (A) are for different brain sections showing areas of increased activity during pain. Reprinted with permission from Petrovic et al.[36] Copyright © 2002 American Association for the Advancement of Science.

Table 8.2
Percentages of Patients Committing or Attempting Suicide

Type of Drug	Suicides (%)	Suicides + Attempts (%)
Placebo antidepressant	0.4	3.1
Old antidepressant	0.7	4.1
New antidepressant	0.8	3.6
Placebo antipsychotic	1.8	5.1
Old antipsychotic	0.9	6.6
New antipsychotic	0.7	5.7

244   were 241 suicides and attempted suicides. Both antidepressant and antipsychotic drugs were investigated in these two studies. As can be seen, there were no statistically significant differences in the rates of suicides and attempts between the placebo and drug groups either for antidepressants or for antipsychotics, though the outcomes for the two placebo groups were slightly better(!) than both the old and new drugs.

It should be noted, however, that antipsychotic drugs (used to treat conditions such as schizophrenia) are in an entirely different category from antidepressants when it comes to the alleviation of symptoms. Despite their lack of reduction in suicides and attempts compared to placebos, antipsychotic medications have proven their value in alleviating psychotic symptoms.[37]

Media Stories about Antidepressants

Following these various studies by Khan and others, a number of media accounts appeared in 2002 suggesting that antidepressants were no more effective than placebos in clinical trials. As might be expected, this was not a message that drug companies would want the U.S. public to hear. Several developments followed that spate of media coverage. Drug companies appear to have stepped up their aggressive marketing efforts noted earlier. Additionally, a new set of stories began to appear in the media suggesting that the earlier-reported studies may not mean quite what they seemed to mean and that antidepressants really do have value beyond placebos, according to various researchers quoted in the stories.

It is worthwhile to take a brief look at some of the arguments given in the second round of media stories suggesting that antidepressants might not be just placebos after all. One speculation is that the size of the placebo effect is getting stronger over time, making it harder for the antidepressants to distinguish themselves from placebos. As one wag has suggested, maybe they're making placebos better than they used to! This increase in placebo effectiveness supposedly is due to more people becoming aware of the utility of antidepressants and having

245 higher expectations of their power. While that suggestion may be true, it would not account for placebos increasing their effectiveness *relative* to active drugs over time. Moreover, there is no clear evidence there has in fact been a steady increase in the magnitude of the placebo effect over time. Given the large variations from study to study, one can always cite two studies— one earlier and one current—to make a claim for such an increase, but no such trend is seen in the studies collectively.

Another argument advanced in the revisionist media stories involves the timescale. It is suggested that the true effect on depression treatment cannot be seen in as short a time as eight weeks, which is how long a typical clinical trial lasts. For example, according to an article in the *Washington Post*:

> A placebo group might show equal improvement to a medicated group at first. But, if a placebo study were done over a period of several years, the placebo group would almost certainly fall behind."[38]

The certainty expressed in the preceding quote is amusing, since there appear to be no studies that test that specific question directly. On the other hand, the implication of the quote might still be true, but perhaps not for the reason its author believes. If a placebo antidepressant were administered over a period of several years, it might in fact lose its power relative to an active drug. But that could simply be attributed to increasing patient awareness that he or she was in the placebo group, based on the absence of any readily observed side effects.[39]

It seems that this may be one of those cases where the original media stories about how antidepressants were probably no more effective than placebos may have been correct, and the second wave of revisionist articles (possibly stimulated by deep drug industry concern) got it wrong. Whatever the truth about the effectiveness of antidepressant drugs, one could take the position that as long as they help people, what is the difference if they are no better than placebos? After all, many people take all sorts of "natural" remedies to treat depression, such as St. John's Wort, and there is every indication they are simply placebos. It is even possible that all of psychotherapy is simply

246 based on a placebo effect, as was suggested earlier. The big difference is that with antidepressants, which are considered to be active drugs, the U.S. government through the Food and Drug Administration (FDA) attests to the fact that the benefits of a drug outweigh the risks and that the drug does have actual benefit beyond a placebo.

The Role of the FDA in Approving New Drugs

Any drug, by definition, contains active ingredients that can have an effect on the body, and it must prove its effectiveness in double-blind clinical trials. Rules governing the FDA criteria for approval can be found in the document "Providing Clinical Evidence of Effectiveness for Human Drug and Biological Products" (www.fda.gov/cder/guidance/1387fnl.pdf) and other documents at the web site www.fda.gov/cder/guidance/index.htm.

For example, the FDA requires that the results of all trials, positive and negative, be turned in before it judges the effectiveness of a new drug. And in most cases it is required that the new drug demonstrate its effectiveness at the 95 percent confidence level in at least two double-blind clinical trials. In some cases only one trial is sufficient if the drug is merely being used in a different dosage or to treat a different (but related) illness than the one for which it was previously approved.

These FDA criteria are not as stringent as they might seem for a number of reasons connected with items that the FDA chooses *not* to regulate about the drug testing process, including the following:

1. *Active placebos.* The FDA does not require that drug companies use active placebos, whose side effects mimic the drug being tested. As a result, double-blind studies can become unblinded and a false statistically significant result can occur rather easily, as explained in the appendix. (The FDA requirement that side effects of drugs be described to patients in clinical trials increases the likelihood of studies becoming unblinded.)

247 **2.** *Effects of unblinding.* The FDA does not require that drug companies discuss the impact of unblinding in their clinical trials and how it might have biased the results. In fact, according to one study, less than 5 percent of the time is the unblinding issue addressed in trials published in the major psychiatric and medical journals.[40]

3. *Effect size.* In general, the FDA does not require that an active drug beat placebos by any specific amount. Only in specific types of drugs, such as antibiotics or beta agonists are specific amounts currently set (10–15 percent). There is no specific similar requirement for antidepressants and most other drugs, which could, in principle, be found to beat placebos by 1 percent and be considered effective. (This last example assumes that the 1 percent difference is statistically significant, which would require an enormous number of patients.)

4. *Negative results.* Although the results of negative trials must be supplied to the FDA, the agency does not prohibit drug companies from having a veto over their publication in the medical literature. In fact, many drug companies now *require* that scientists conducting the trials sign agreements stipulating that the company has the right to veto the publication of results. This deplorable stipulation means that the published literature has become biased by an unknown amount. Although researchers wishing to do a meta-analysis of the effectiveness of a drug can still rely on the completeness of FDA archives, they could not do an unbiased analysis using the published literature, given the absence of many negative results. Doctors wishing merely to stay current on what drugs to prescribe for patients are unlikely to wade through the FDA archives.[41] Even more to the point, the practice of allowing a company veto over publication can cost lives. In one instance under litigation, cancer researchers violated their no publication agreement with a drug company and were sued as a result. The researchers published the results because they believed that the drug was killing patients and they were unwilling to suppress the data, even though they had signed a no-publication agreement.

248 5. *Inconclusive results.* In addition to the categories of positive and negative results for a given trial, the FDA recognizes a third category of inconclusive or "no test" results, which do not count! For example, let us suppose that you did a study comparing a new drug against a placebo with another control group using an old drug. Let's say you found that neither the new drug nor the old drug outperformed the placebo. In that case, you couldn't be sure whether that result was because the new (and old) drugs were both ineffective, or because your placebo group had a better than expected outcome, just by chance. Interestingly, the FDA would count that as a "no test" result, not a negative one, and it would not count against the evaluation of the new drug.

While writing this chapter I contacted the appropriate official at the FDA (Robert Temple, Associate Director for Medical Policy at the FDA's Center for Drug Evaluation and Research) to learn how many "no test" results a new drug could have and still receive approval. I had asked, by way of example, if a new drug could have as many as 18 null (no test) results out of 20 trials. A portion of Dr. Temple's response is shown below, and his full response is shown in a footnote:

> I suppose that if you had 18 "no test" studies and only 2 that had assay sensitivity (ability to tell active drug from placebo) one might consider that adequate, but that has never happened.[42]

In other words, a drug could conceivably pass muster if it had two positive trials and 18 trials all with inconclusive ("no test") results.

Why the "No Test" Category Is Questionable

At first thought it may seem reasonable to have a "no test" category of results in addition to positive and negative ones. Such a description might seem appropriate for the outcome of a trial in which both a proven old drug and an unproven new one failed to beat a placebo. This outcome

249 could indeed come about because by chance the placebo group did much better than expected. However, if the "proven" old drug really was more effective than a placebo at the usual level of statistical significance, such an outcome should happen by chance no more than once in 40 trials on average. The chance of accumulating N "no test" results in succession is $1/40^N$. If a company seeking approval of a new drug had one "no test" result along with two positive ones, that might not be a matter for concern. But, given the FDA rules of not counting "no test" results, a company could accumulate a large number of them—conceivably as many as 18. Such an outcome could happen by chance one time in 40^{18} or one in 70 billion billion billion—assuming that the placebo group was extraordinarily lucky trial after trial against a truly effective old drug. (The chances of that happening are much less than that of most life on Earth being extinguished in the next second by a large meteorite impact.) A much more plausible interpretation of such an unlikely outcome is that the old drug was in fact no better than a placebo, despite its having received prior FDA approval. This raises the suspicion that the "no test" category is a useful fiction, which allows the FDA to avoid revisiting the question of the effectiveness of older "proven" drugs that don't seem to be able to beat placebos in trials of new drugs.

The FDA is an agency with many dedicated and sincere individuals. However, an objective look at the overall picture of the FDA regulations concerning the approval of new drugs might lead one to conclude that the agency merely wanted to preserve the form of controlled double-blinded studies, while undermining the real intent—the unbiased evaluation of he effectiveness of new drugs. Obviously, some of the FDA requirements that lead to problems of bias (through unblinding), such as informing patients of drug side-effects can be justified on ethical grounds. But the unblinding problem has a simple policy solution: either requiring active placebos be used when they are available and can be administered without harm to patients, or requiring that the effects of unblinding be discussed when trial results are reported, and taken into account when doing statistical analysis.

250   One therapist, who believes strongly that antidepressants are not merely placebos, has expressed the view that even if they were, it could be harmful to patients if such a belief were to become widely accepted.[43] In such a case, the placebo benefit presumably would lessen, because placebos don't work as well if you believe they are not real remedies. That therapist's concern has some validity, but an acceptance of the view that antidepressants are merely placebos is so far from being widespread among physicians that his concern seems unwarranted. Moreover, there are important advantages to disseminating the truth about the effectiveness of mental health and other remedies. Such widespread dissemination would promote the search for remedies that have genuine value, beyond the placebo effect, and add pressure to change FDA rules, so that both the true spirit as well as the formality of the double-blind method is followed.

The Nocebo Effect

The nocebo (or anti-placebo) effect is less well-known than the placebo effect, and could be called its "evil twin." In the placebo effect you expect to improve as a result of some treatment, and you later do improve. With the nocebo effect, you expect a poor outcome, and you later get one. ("Think sick, become sick.") Nocebos are usually not administered intentionally by physicians, unless they are unusually sadistic, but they may be inherent in the messages given by some physicians who lack a caring attitude or are unable even to feign one. Much of the data on the placebo effect has been accumulated in clinical trials, where placebos are used as controls. Since nocebos have no such usage, we would expect that data on them are much sparser. In fact, a search of the Medline database for the keyword "nocebo" gave a mere 33 hits, compared to over 70,000 hits for the keyword "placebo." (That comparison should not be taken too literally, because some researchers writing about a negative placebo effect might not use the word "nocebo.")

251   Ethics represents another barrier to learning about the effects of nocebos, at least for experiments with humans, although that was not always the case. In one interesting century-old experiment, an allergy sufferer was set wheezing when shown an artificial rose. In another experiment from the 1980s, a group of college students were told that when a small electric current was passed through their heads they might get a headache. Two-thirds of the students reported headaches, even though no electric current was actually used. In another similar experiment, reminiscent of the cartoon of Dr. Nocebo (figure 8.4), a group of hospital patients were given ordinary sugar water and told it was an emetic that would induce vomiting. Eighty percent of the patients actually did vomit as a result.

A number of "natural" experiments also attest to the importance of the nocebo effect, or the power of negative suggestion. As part of the Framingham Heart Study a group of middle-aged women were studied for 20 years for their incidence of heart attacks and deaths resulting from them. It was found that women who considered themselves likely to die of coronary heart disease were 3.7 times as likely to do so as those who did not consider themselves to be at risk, even after the two groups were matched for known risk factors.

A particularly interesting natural nocebo experiment has been called the "Hound of the Baskervilles effect" after a famous Sherlock Holmes story in which Charles Baskerville dies from a stress-induced heart attack.[44] The study was based on

You have six months to live   I want a second opinion   OK, you're ugly

Figure 8.4 Dr. Nocebo sees a patient.

252 noticing a particular linguistic coincidence. In Cantonese, Mandarin, and Japanese the words for "four" and "death" are pronounced nearly the same. As a result, many Chinese and Japanese people believe the number four to be very unlucky. David Phillips and colleagues examined the daily mortality rates for over 200,000 Chinese and Japanese American deaths between 1973 and 1998. In comparing the deaths from chronic heart disease on different days of each month, it was found that on the fourth day, rates were 13 percent higher than other days—a statistically significant amount.

In California, where superstitious fears might be more likely to be reinforced by larger Asian population concentrations, the effect was even higher: 27 percent above the average of other days. No comparable fourth day peak in heart attack mortality was observed in the mortality records for Caucasian Americans. Phillips and colleagues were able to rule out various causes of this fourth day peak associated with changes in diet, exercise, alcohol intake, or meditation regimes, and they concluded that the observed increase in heart attack mortality was due to psychological stress.

It is possible that effects such as those as described in Phillip's study can occur even in healthy individuals. According to Herbert Benson, a physician who has studied the placebo and nocebo effects extensively, animal studies show that an area of the brain can cause ventricular fibrillation when it is stimulated, and some researchers believe that extreme fear, even in a dream, initiates such a process, leading to death.[45] One extreme example of being literally "scared to death" may involve voodoo hexes. "Voodoo death" was first described in a classic 1942 paper by physiologist Walter Cannon who had heard many stories of people in primitive societies who apparently died from fright after being subject to a voodoo curse or hex in which a bone is typically pointed at the victim.[46] Cannon came to this conclusion after attempting to rule out foul play or other causes of death. Although his conclusion necessarily remains speculative, Cannon's account of the typical victim is most haunting: "He stands aghast, with his eyes staring at the treacherous pointer,

253 with his hands lifted as though to ward off the lethal medium, which he imagines is pouring into his body. His cheeks blanch, and his eyes become glassy and the expression on his face becomes horribly distorted."

Voodoo death aside, it is clear that stress and other psychological factors can adversely affect health through a connection with the immune system. In fact, family physicians estimate that two-thirds of the illnesses they treat are caused by psychological factors or stress. Some of the life-threatening conditions to which stress has been linked as a contributing factor include heart disease, strokes, cancer, breathing problems, accidents, suicide, and cirrhosis of the liver. Studies have shown that stress affects circulating levels of cortisol, one of the chemical messengers released by the adrenal gland. Elevated cortisol levels can impair the number and function of immune cells and leave you more vulnerable to disease. A similar chemical mechanism appears to occur in the reverse direction in the case of the placebo effect, which is accompanied by the elevated release of neurotransmitters such as endorphins that have benefit in pain reduction. Neither of these observations can be said to give a definitive explanation of the placebo and nocebo responses, since it is not understood how the power of suggestion induces these particular chemical releases.

Stress can induce nocebo responses in a range of different settings. One study of highly stressed caregivers of Alzheimer patients, for example, found that wound healing took 24 percent longer in the caregivers than in people who were not stressed. The pressures of daily life have can lead to a host of ills, including elevated levels of depression and anxiety, particularly for people living in geographic areas considered likely to be the target of future terrorist attacks. The dubious practice of declaring terrorism alerts without providing information about specific ways to avoid danger can only exacerbate stress levels in vulnerable individuals, according to some psychiatrists.

Instances of mass hysteria often occur in schools or workplaces involving repetitive or boring work, giving rise to the name "assembly line hysteria." A few people may fall ill and

254 experience such symptoms as fainting, pain, nausea, and headaches, and many others who observe them begin to experience the same conditions. The psychological impact of the distress of others on our own health need not involve a direct observation. It is well known that instances of suicide or suicide attempts among teenagers are sometimes the result of a prior suicide of a famous person or a schoolmate.

Radiation Phobia — The Ultimate Nocebo?

Nuclear or ionizing radiation is, of course, harmful to living things. At high doses it can kill, while at lower doses it can cause cancer and genetic damage. Many scientists believe that there is no radiation dose below which radiation is harmless. However, there is a minority view that low doses may actually be beneficial—see my book *Nine Crazy Ideas in Science.*[47] In many respects nuclear radiation or perhaps radiation phobia is the ultimate nocebo. First, radiation cannot be seen or smelled, and so the level present is simply left to your imagination unless you are equipped with a detector. Second, radiation is known to have harmful, even lethal effects, and those effects are believed to be present to some degree even at low doses, according to conventional wisdom. Third, the world has already witnessed the horrible consequences of large-scale radiation effects on humans as a result of the World War II Hiroshima and Nagasaki bombings and the nuclear reactor disaster in Chernobyl.[48] In contrast, the worst nuclear power plant accident in the United States, at Three Mile Island, released an amount of radiation to the environment that probably caused zero direct illnesses, but a great deal of psychologically induced (nocebo) illnesses among citizens in the area. Excessive fear of radiation lives on, and probably explains why the notion of a "dirty bomb" (a conventional bomb that would disperse surrounding radioactive material) remains high on the list of terrorist threats. In contrast to a nuclear bomb, a dirty bomb, of the sort we would most likely encounter, would probably contaminate surrounding areas only slightly, and would expose most people to levels of radiation that would elevate their cancer risk by only a very tiny amount. Nevertheless, the ensuing panic could cause a great deal of harm.

255   Pending or prospective litigation can also be a major nocebo. If you have received an injury in an automobile accident that was someone else's fault, you have a strong incentive not to recover before the insurance company settles the case. Even if you are not faking injury, this economic incentive can still exert psychological pressure that impairs your recovery. The prospects of recovering large sums of money may even make you believe you are sick, when nothing may be wrong with you, apart from the usual aches and pains we all experience.

Some observers believe that silicone breast implants are another example of this type of nocebo. For 30 years women were apparently happy with these implants and reported few problems. Following a spate of media reports about the horrors of implants, many women began suffering a variety of strange illnesses that they attributed to the implants. Although no reputable scientific study has confirmed that women with silicone implants experience more serious illness than other women, this absence of scientific evidence has had little impact on jury verdicts and public opinion. When law firms trumpet in advertisements that "$100,000 OR MORE MAY BE OWED TO YOU" if you have breast implants, it is hard to resist the conclusion that the prospect of financial gain can act as a powerful nocebo.[49]

But we should not conclude that nocebo and placebo effects operate at a conscious level, or even that they require your belief in order to be effective. Experiment with animals, for example, show that they can also experience placebo and nocebo effects. In one study rats were periodically injected with an immunosuppressive drug and simultaneously fed saccharin-flavored water.[50] After some time, the rats continued to be fed the flavored water but now without the drug. These rats (and a second group not fed saccharine) were later injected with bacteria that challenged their immune systems. The rats that had been conditioned to associate saccharin-flavored water with the immunosuppressive drug had significantly lower levels of antibodies to the bacteria than the other group, even though they no longer received the drug. In this case, the association that the rats formed between the saccharin solution and the harmful drug appears to have "taught" them to experience the

256 effects of the drug on their immune system for many days after its use was discontinued. The flavored water could well be thought of as a nocebo in this instance.

Other experiments using other drugs administered to animals further illustrate the importance of conditioning in creating placebo and nocebo responses. In one experiment with dogs, conditioning was able to enhance the effect of morphine. Even more surprising, sometimes conditioning can reverse the effect of a drug in such studies—causing, for example, an increase in activity following a period of conditioning with a tranquilizer, or causing hyperglycemia by means of inert drugs following a period of conditioning with insulin injections.

Conditioning, of course, occurs in humans as well as animals. But explanations of the placebo and nocebo effects that put the entire focus on conditioning are too simplistic. All the factors discussed previously, including conscious expectation, subconscious association, and natural regression to the mean, probably play a role. If we take a placebo pill when our pain is at its worst, for example, and then find that we feel better because of a spontaneous recovery (regression to the mean), we become conditioned to associate the pill with relief of pain (conditioning).

The nocebo and placebo effects have an interesting, but not entirely symmetrical relationship. For example, as shown by H&G's study, regression to the mean plays a significant role in the placebo effect. But spontaneous variation of symptoms would play no role in explaining the kinds of nocebo phenomena discussed above. So those who would maintain that the placebo effect is essentially just a matter of regression to the mean, with little or no role for conscious expectation and conditioning, would not be able to account for nocebo phenomena. And, by extension, if negative expectations and conditioning can affect our health, it seems quite plausible that positive ones should do likewise.

Placebos undoubtedly have a major impact on our well-being generally—in some cases perhaps as much as conventional "proven" remedies—even in matters of life and death.

257 Many doctors may not be overly fond of the placebo effect. They would much prefer to rely on scientific medicine than bedside manner and sugar pills. But, like it or not, the placebo effect can be a formidable ally in promoting healing—and the nocebo effect in hindering it.

Even self-administration of placebo remedies can have important value on some occasions. Relying on the placebo value of quack cures can be harmful if it keeps us from seeking proven treatments. But if you have money to burn and you want to take some advertised remedy that is said to "promote intimacy" or "improve your concentration," what's the harm? And I hear that wearing a copper bracelet can do wonders for your golf game, too. Just stay away from those nasty nocebos! I suppose I'd give the idea that a dummy sugar pill can cure you (or make you sick) as 0 flakes.

Appendix: *Creating a Spurious Statistically Significant Result from Unblinding*

Here we consider a simple mathematical model to show how a drug that is no more effective than a placebo can yield a false statistically significant benefit in a clinical trial as a result of unblinding. A contrary opinion is provided by Quitkin in a 2000 article.[51] Quitkin argued that the lack of any significant difference between the average response rate or effect size in trials with inert and active placebos indicates the irrelevance of inert placebos. However, as our model will show, there can be a significant difference between the effect sizes in the active drug and placebo groups, depending on the extent of unblinding in a given trial. The basic point of the model is that the size of the placebo effect is dependent on whether people become unblinded, and the fraction of people becoming unblinded is different in the active drug and placebo groups.

After describing the model and some numerical results, various potential limitations will be considered and shown to be unimportant. The model assumes for simplicity a head-to-head

258 comparison between a drug and a placebo, with no third comparison group. The following parameters are used in the model:

n = the number of participants in each group. Groups are assumed to be equal size.

f = the fraction of the placebo group patients who become unblinded.

p = the fraction of blinded people who report a reduction of symptoms, i.e., the size of the placebo effect.

r = the reduction of the placebo effect among unblinded individuals who take placebos. For example, if $r = 0.3$, and $p = 0.6$, then if you correctly deduce that you are taking placebos, your chances of a reduction of symptoms would be 42 percent, i.e., 30 percent less than 60 percent.

P_{eff} = net placebo effect for the placebo group (blinded plus unblinded). This quantity is what is usually measured in a clinical trial, rather than p. It is given by $P_{\text{eff}} = p(1 - rf)$.

A randomly chosen individual in the "active" drug group will experience a reduction of symptoms with a probability p. If

Table 8.3

Means and Variances for Different Groups

Distribution	Mean	Variance
Active drug group	$N_1 = pn$	$\sigma_1^2 = np(1 - p)$
(a) Placebo group (unblinded)	$(1 - r)fnp$	$\sigma_u^2 = np(1 - f)(1 - p)$
(b) Placebo group (blinded)	$(1 - f)np$	$\sigma_b^2 = npf(1 - r)(1 - p + pr)$
(a) + (b) Entire placebo group	$N_2 = (1 - r)pfn + (1 - f)np$	$\sigma_2^2 = \sigma_a^2 + \sigma_b^2$
Effect size: active drug minus placebo	$N_1 - N_2 = pfnr$	$\sigma^2 = \sigma_1^2 + \sigma_2^2$ $= np[2 - 2p + fr(2p - 1) - fpr^2]$

259 many trials are conducted, the number of persons in the active
drug group that report a reduction of symptoms can therefore
be assumed to follow a binomial distribution. The placebo
group consists of two subgroups: (a) the unblinded group, and
(b) the group that remains blinded. Each of these subgroups
has its own binomial distribution. The distribution of persons
receiving a reduction of symptoms in the placebo group is a
convolution of these two distributions for groups a and b. The
means and variances of all the respective binomial distributions are given in table 8.3, along with that of the "effect size,"
i.e., the difference between the active drug and placebo groups.

For statistical significance at the 95 percent confidence level
(two tails), we use $t^2 = (N_1 - N_2)^2/\sigma^2 \cong 4$. Using the last row of
table 8.3, we obtain a quadratic equation in r: $Ar^2 + Br + C = 0$
with $A = pnf^2 + 4pf$, $B = f - 2pf$, and $C = 8p - 8$. The value of r
from the solution of this equation tells us how likely it is that
unblinding can produce a false-positive drug efficacy. (For example, if we found $r = 0.25$ it would mean that if unblinded individuals experience a placebo effect that is at least 25 percent
less than its value among those who remain blinded, then the
trial will yield a false-positive result due to unblinding.)

Results of Model. The likelihood of unblinding yielding a false
statistically significant result in any given trial depends on
whether the observed reduction in the placebo effect for unblinded individuals exceeds the value r computed from the
preceding quadratic equation. The r values so computed depend on the values of the three "input" parameters, p, n, f.
Table 8.4 shows how r depends on eight combinations of values
of these three input parameters. Also shown is the calculated
value for p_{eff}. In general, we see that, as expected, smaller r values result, the larger we make n, p, and f. The eight cases are or-
dered from left to right in order of increasing r—meaning a decreasing likelihood of a false statistically significant result. For
example, case 1, which has the smallest r and is therefore most
likely to yield a false statistically significant result, has $n = 80$,
$p = 0.6$, and $f = 0.8$, which yields $r = 0.32$. In this case, a spuri-

260 ous statistically significant result would therefore arise as long as individuals who become unblinded experience a placebo effect at least 32 percent less than that of those who remain blinded, which seems quite plausible in practice. Case 8—least likely to yield a false statistically significant result—has $r = 0.69$. For this most conservative case unblinding would occur when the size of the net placebo effect is only 29 percent. Hence, most of the eight choices of the input parameters yield potentially realistic r and p_{eff} values corresponding to false-positive results from unblinding.

Potential Limitations of the Model

1. *Only two groups.* Some drug trials involve three groups, not two. The model applies equally well to that case, however, because the two active drug groups, if they experience the same degree of unblinding, would show the same spurious statistical significance over the placebo group according to the model. (It is reasonable to assume a similar degree of unblinding in the two active drug groups if the new and old drugs have similar easily recognizable side effects and the placebo has none.)

2. *Partially active drugs.* In the model we assume that the "active" drug is no more effective than a placebo, because the goal was to show how a statistically significant result

Table 8.4

Calculated r and p_{eff} Values for Eight Possible Combinations of n, p, and f

	Case 1	Case 2	Case 3	Case 4	Case 5	Case 6	Case 7	Case 8
$n =$	80	80	80	40	80	40	40	40
$p =$	0.60	0.50	0.60	0.60	0.50	0.50	0.60	0.50
$f =$	0.80	0.80	0.60	0.80	0.60	0.80	0.60	0.60
$r =$	0.32	0.38	0.42	0.43	0.51	0.53	0.57	0.69
$p_{eff} =$	0.45	0.35	0.45	0.39	0.35	0.29	0.40	0.29

Note. The eight cases are listed in decreasing likelihood of a false positive result, i.e., increasing size of r.

261   could occur even in this extreme case. We might have assumed a small, but not statistically significant benefit to
the active drug over a placebo. The assumption in the
model was therefore a conservative one.

3. *Unblinding of the physician or trial assessor.* In the model we
   took into account the unblinding of the patient, but not
   the physician, which could be an additional source of bias
   leading to a spurious statistically significant result.
   Ignoring the unblinding of the physician, which has been
   shown to occur in clinical trials, is therefore also a conservative assumption.

4. *Increase in placebo effect.* It is possible that when patients in
   the active drug group become unblinded the size of the
   placebo effect increases for them. Such an increase due to
   unblinding has been ignored in the model, but this again
   is a conservative assumption that would not affect the
   conclusions.

5. *Variable* r. It is possible that the extent of reduction in the
   placebo effect is correlated with the extent to which someone becomes unblinded. People who are quite certain
   they are being given placebos might experience more of a
   reduction in benefit than people who merely suspect that
   may be the case. This is accounted for in the present
   model if r is understood to represent the average reduction in the placebo group among unblinded individuals.

6. *Realism of input parameters.* As already noted, the case
   showing the most significant result (smallest r) was for
   the largest n, p, and f. No information on the parameter p
   (the size of the placebo effect among individuals who remain completely blinded) could be found in the literature. However, published information does exist on p_{eff}.
   The value for p_{eff} in each of the eight example cases given
   in the table are consistent with widely varying values
   seen in actual clinical trials of antidepressants. For example, in one meta-analysis by Arif Khan, the average $p_{eff} =$
   0.3.[52] The unblinding fraction seen in clinical trials of antidepressants is typically around $f = 0.8$, but the value for

those in the placebo group may be only 0.6.[53] Both of these values have been used in the table.

7. *Realism of calculated* r *values.* A search of the relevant literature revealed no published values for *r*, the average reduction in the placebo effect for unblinded individuals. Until an experiment is done we therefore do not know for certain if the values used in table 8.4 are realistic. The *r* value for the most favorable case ($r = 0.32$ does not seem at all unreasonable, although the least favorable value listed ($r = 0.69$) might be.

Conclusion. The results of the model suggest that unblinding can explain the apparent efficacy of some drugs (especially antidepressants) in clinical trials, even if they are in fact no better than placebos, as previously suggested in several earlier meta-analyses.[54] To confirm the analysis presented here, it would be useful to experimentally investigate the value of *r* and *f* in a large clinical trial to confirm that the values used in the model are realistic. An alternative use of the model would be to correct for unblinding effects prior to assessing statistical significance. This correction could be done by using the trial data to obtain values for *r*, *f*, and *p*, and then performing a *t* test. Clearly, greater attention needs to be paid to methods to prevent unblinding, such as active placebos (when ethically justified), and a higher level of statistical significance than $p < 0.05$ needs to be used when unblinding is a real possibility.[55] This model also highlights our ignorance about the mechanics and components of the placebo response.

9 Should You Worry about Your Cholesterol?

UFFE RAVNSKOV is a man on a mission. He wants to convince you that with very rare exceptions you shouldn't worry about your cholesterol. Ravnskov is a Swedish physician, now retired from private practice, who has explained his ideas in a book: *The Cholesterol Myths: Exposing the Fallacy That Saturated Fat and Cholesterol Cause Heart Disease.*[1] Additional information about Ravnskov and his controversial ideas about cholesterol can be found at the web site www.ravnskov.nu/cholesterol. Information about his career and his publications can be found at: www.rondellen.net/ravnskov_eng.htm.

Cholesterol: The Good, the Bad, and the Ugly

Before considering the merits of Ravnskov's ideas, it's worth remembering that whatever the possible harmful effects of cholesterol might be, the substance is essential for normal body functioning. Cholesterol is a hard waxy substance that is present in animal fats, dairy products, and especially eggs. Even if our diet included no cholesterol, our body would produce about a gram of it each day for a variety of important functions. Within limits, the more cholesterol we eat, the less our body needs to manufacture, and vice versa. Thus, a significant portion of your cholesterol level is endogenous, or independent of outside factors, such as your diet.

Bodily uses of cholesterol include a basic construction material of cell membranes (which would otherwise fall apart) and the manufacture of sex hormones, vitamin D, bile (needed to absorb fat in the intestines), and myelin (the fatty coating surrounding nerves). Cholesterol is also secreted by glands in the

264   skin, where it serves as a protective barrier to dehydration and infection. Clearly, cholesterol is an essential substance in the human body. In addition, although the cause remains uncertain, very low levels of cholesterol have been identified as a risk factor in a variety of illnesses, and they can shorten your life span significantly.[2] More than 20 studies have been done on an association between very low cholesterol levels and cancer, and most report there is a link, though it is unclear whether low cholesterol is the cause or the effect of cancer.[3] According to one study, if your cholesterol level falls in the lowest 5 percent of the population, then you have an elevated risk of death from all causes of around 14 percent if you're male and 66 percent if you're female.[4]

So much for the good side of cholesterol. Most people's familiarity with cholesterol relates to the problems associated with excessive levels of the substance. The so-called "heart–diet" theory holds that diet is the leading cause of coronary heart disease, which is in turn the leading cause of death in many countries, including the United States. In his book, Ravnskov tries to debunk this theory by discussing a series of "cholesterol myths." In particular, he disputes the idea that excessive dietary intake of saturated fats and cholesterol is a major cause of heart disease. Unlike cholesterol, saturated fats are found in foods from both animal and plant sources. Although saturated fats from plant sources, such as coconut oil, don't contain cholesterol, they may stimulate the body to produce it.

How Does Coronary Heart Disease Develop?

In the conventional heart–diet theory, the development of coronary heart disease (CHD) takes place in a three-step process. First, we develop high levels of cholesterol in the blood, as a result of a bad diet that is too rich in cholesterol and saturated fats. Second, the high cholesterol in the blood causes the development of "plaque," which blocks arteries to the heart. The third and final step is the development of CHD due to the blocked arteries.

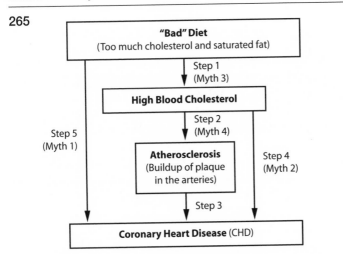

Figure 9.1 The five steps in the heart–diet idea, and four of Uffe Ravnskov's "myths." Picture is from Ravnskov,[1] printed with permission.

The three steps in the development of CHD are shown in figure 9.1 as the arrows at the center of the drawing taken from Ravnskov's book. The figure also shows some of his specific challenges to the heart–diet idea, i.e., his "myths." For example, "myth 1" is Ravnskov's claim that there is no real evidence showing that a diet high in cholesterol and saturated fat causes coronary heart disease. The only step in the three-step process that Ravnskov doesn't challenge is that blocked arteries cause CHD. On the other hand, while blocked arteries do mark the onset of CHD, the full progression does not occur in all cases. If the blockage of an artery occurs gradually enough, alternate pathways around it can be created when other coronary arteries widen and allow adequate blood flow to the heart to be maintained. In effect, the body performs a heart bypass operation on itself!

Although the analogy might seem appealing intuitively, the blockage or "hardening" of arteries, also known as atherosclerosis, is not exactly like the gradual buildup of sludge in a sewer pipe or a sink drain. Instead, the process appears to be initiated by an injury to the inner wall of the artery that roughens it. As a result, cholesterol particles begin to adhere to the

266 wall at the injured site. Other changes occur as part of an inflammation reaction in which hard calcium deposits and fi-
brous material are added to the wall, so as to create a smooth
area that allows blood to flow more easily past the injured site.
As the lesion known as plaque gradually grows in size and the
arterial wall thickens, there is less cross-sectional area for blood
to flow past the obstruction. The narrowed artery can cut off
blood flow to the heart sufficiently to cause a heart attack,
which occurs when a portion of the heart muscle is deprived of
oxygen and dies. In a less severe case, CHD can manifest itself
through angina, which is the chest pain you feel, particularly
when exercising, because the heart cannot get enough blood,
due to the narrowed coronary arteries.

Another potentially fatal possibility can occur if the plaque
ruptures and forms a blood clot. In this case, if the blood clot
reaches the brain, the result is likely to be a stroke. A short animated sequence of the formation of plaque on the inner wall of
an artery, starting with an initial injury and ending with the release of a blood clot (thrombus) to the bloodstream, can be
found at the web site www.medimagery.com/atherosclerotic.
html. Ravnskov doesn't question the process just described,
which leads to the development of blocked arteries, but he does
question the role of diet and high cholesterol levels in the blood
as being the relevant causes. (It might be noted here that no research exists to show that arteries can become blocked from
cholesterol building up on a healthy uninjured arterial wall,
similar to the way a sewer pipe might become clogged.[5])

Initial Impressions of Ravnskov's Ideas

I first learned of Uffe Ravnskov's idea that cholesterol is harmless only a few months ago, when I attended his lecture in
Washington, DC. For me, it was love at first sight! Not that I
had any immediate attraction to Dr. Ravnskov, or even his
ideas, but I knew instantly that it would become the topic of a
chapter of this book. The idea that cholesterol was harmless
had everything I was looking for: it was highly controversial,

267 was relevant to the concerns of millions of citizens, and had many data that could either support or refute it.

Although I was initially drawn to Ravnskov's idea because of the very impressive amount of data that he presented in his Washington lecture, I also had some early qualms about it. In part, these qualms revolved around his claim that most researchers have systematically ignored or downplayed contrary evidence, which keeps other scientists and the public ignorant about the ideas he presents. For example, on his web site Ravsnkov notes that: "Many of these facts have been presented in scientific journals and books for decades but are rarely told to the public by the proponents of the diet–heart idea. The reason why laymen, doctors and most scientists have been misled is because opposing and disagreeing results are systematically ignored or misquoted in the scientific press."[6]

Without casting any aspersions on Ravnskov, it is a fact that cranks often claim that there is much evidence for their views that the scientific establishment has simply chosen to ignore. It didn't help any that Ravnskov put "M.D., Ph.D." following his name on his book's cover, as if to reassure the reader that he really was not a crank. Nor did I find it reassuring that his book was published by an independent publisher. In fact, Ravnskov notes that the book was turned down by other publishers and literary agents, because it was considered to have "no commercial interest." (How could a book on such an important subject fit that description, unless it was nonsense?) Equally worrisome, the book has not been reviewed in any major scientific or medical journal since its publication.[7] By his own admission, Ravnskov's challenge to the heart–diet theory has been virtually ignored by the medical establishment. (On the other hand, in fairness to Ravnskov, while his is definitely a minority position, he is by no means alone in questioning the role played by cholesterol in the development of coronary heart disease.)

The very preposterous flavor of Rasnkov's ideas also made me suspicious. It was reminiscent of the theory that HIV is an innocent virus and not the cause of AIDS, which I analyzed critically in *Nine Crazy Ideas in Science*.[8] As with Peter Duesberg, the primary supporter of that highly controversial view on

268 AIDS, Ravnskov seems to believe 100 percent in everything he writes and appears not to entertain any doubts about it. That kind of firm attachment to one's theory can either indicate a great preponderance of evidence in favor of the theory or a less than critical judgment of contrary evidence.

In addition, although he has an extensive publication record in the medical literature, Ravnskov admits to not having been directly involved in heart disease research. In several respects, however, this lack of direct research involvement may be an asset. It means that one is less tied to traditional ideas and freer to think "outside the box." More importantly, unlike those who do research involving cholesterol, Ravnskov is not beholden to commercial interests. This is an important consideration in an era when medical researchers must depend on funding from drug companies to pay for expensive trials of new drugs, which usually require at least a million dollars.[9]

Although I found still more reasons to be suspicious of Ravnskov's ideas once I started reading his book, I also had a strong reason to want to believe it as well. The month before Rasnkov's lecture I had stopped taking the medicine to keep my cholesterol level down and also decided that I would no longer stay away from "bad" foods that I really liked. Life is too short to live like that. Ravnskov's claims, if true, would add an element of rationality to my previous pair of decisions. So much for the "full disclosure" concerning my initial biases pro and con regarding this topic. My initial biases may not be relevant anyway, since I wound up changing my mind at least three times about Ravnskov's controversial ideas as I researched this chapter. In what follows, let us take a close look at the various claims he makes and see how good the evidence is in each case.

Do High-Fat Foods Cause Heart Disease ("Myth 1")?

The idea that foods high in fat are the cause of heart disease is the first "myth" that Ravnskov attempts to refute in his book.

269 The graph shown in figure 9.2 reproduced from his book is instructive in this regard. This plot shows how age-adjusted death rates for coronary heart disease are correlated with percentage fat in the diet for 22 countries. As can be seen in the graph, there is a weak correlation between the two variables, which have a correlation coefficient of 0.39.[10] The correlation between dietary fat and CHD death in figure 9.2 would appear to be even weaker without the drawn curve, which was constructed so as to fit the data for the six countries having labeled

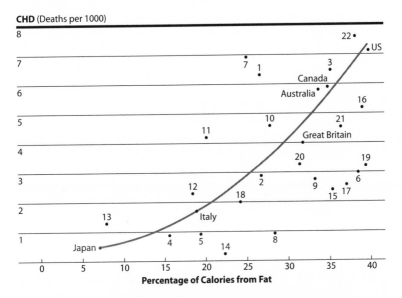

Figure 9.2 Correlation between total fat consumption as a percentage of total calorie consumption, and age-adjusted mortality from coronary heart disease for 22 countries. The six labeled countries were chosen by Keys. The countries are 1, Australia; 2. Italy; 3, Canada; 4, Ceylon; 5, Chile; 6, Denmark; 7, Finland; 8, France; 9, West Germany; 10, Ireland; 11, Israel; 12. Italy; 13, Japan; 14, Mexico; 15, Holland; 16, New Zealand; 17, Norway; 18, Portugal; 19, Sweden; 20, Switzerland; 21, Great Britain; 22, United States. Graph is from Ravnskov,[1] printed with permission. The double appearance of Italy is an error that appeared in Ravnskov. The data used are said to be from reference in note 10, but neither of the two Italy data points appears to agree with that source.

270 points in the figure. (Unfortunately, data points for those six countries appear twice in the figure, since the values used were slightly different from those of Ravnskov.) The only thing special about those six countries is that they were the ones used by physician Ancel Keys in 1953 to make his case that dietary fat was the cause of coronary heart disease. Since data on all 22 nations were available at the time Keys presented his case, his use of those particular six nations shows that his desire to prove his argument appears to have gotten ahead of what the data actually show. As can be seen, when the data for all 22 countries are examined, one finds significant differences in the CHD death rate for any given percentage calories from fat, indicating that the correlation between the two variables is rather weak.

What Is It about the Japanese?

Of all the 22 nations in figure 9.2, Japan has the lowest risk of coronary heart disease and also the lowest percent of dietary fat, apparently making it exhibit A for the heart–diet theory. But might there be a nondietary explanation for the low incidence of Japanese CHD? A 1976 study compared nearly 4000 Japanese-American men living in California with men living in Japan.[11] The Japanese-Americans had much higher rates of CHD. But there were many differences between the two groups besides diet, including many of the usual risk factors for coronary heart disease.

The study author wanted to see if cultural differences might explain some fraction of the higher CHD risk. He found that those Japanese-Americans who became accustomed to the American way of life had a three- to fivefold excess of CHD compared to the group that was least acculturated to the Western lifestyle. The latter group had a similar CHD risk to Japanese living in Japan. This researcher claimed that the differences in CHD risk between the most and least acculturated groups could not be explained solely by the usual coronary risk factors, but some aspect of Japanese culture seemed to be an independent additional variable. In fact, as Ravnskov notes: "The most striking aspect of Dr. Marmot's findings was that *immigrants who became acultured to the*

271 *American way of life, but preferred Japanese food, had coronary disease twice as often as those who maintained Japanese traditions but preferred high fat American food."*[12]

The idea that some aspect of culture is more important than diet as a cause of CHD is a very provocative and interesting finding. But it is undermined by a candid admission by Marmot at the end of his publication that his study had a sizable nonresponse rate. He further notes that his results might be due to selection bias, if those traditional Japanese men among the nonrespondents had much higher rates of CHD. This possibility cannot be dismissed, because "in Japan it is considered shameful to die of a heart attack."[13]

The provocative idea that it is Japanese culture not their diet that is responsible for the low rate of CHD has been further undermined by a more recent study of Japanese men living in Hawaii that Ravnskov fails to mention. The Hawaii group, like the Californians, was found to have elevated risk for CHD two to three times that of men in Japan, but the differences were entirely accounted for by the usual risk factors of coronary heart disease, including smoking, obesity, blood pressure, and alcohol intake. In particular, there was no evidence of any culturally based risk, independent of the known risk factors.[14]

What are we to make of the data on the 22 countries plotted in figure 9.2? To the extent that the data are reliable, and Ravnskov suggests several reasons they may not be, the lesson is that there are many other variables besides dietary fat that must be accounted for to explain the variations seen in the CHD death rates from nation to nation. This observation probably should not come as a major surprise to you. You are probably aware that apart from your diet, your chances of dying from coronary heart disease depend on your age, blood pressure, stress level, smoking status, obesity, family history, exercise habits, alcohol consumption, and whether or not you have diabetes. As noted in the last chapter, even *thinking* that you may be at risk of a heart attack can be a major "nocebo." In addition, although it is not normally considered as such, lack of

272 exposure to the sun may be yet another possible risk factor for CHD, according to some studies.[15]

The large variations seen in figure 9.2 in the CHD death rates between nations are also seen when we look at data *within* a given nation. In a particularly startling comparison, Ravnskov cites a study by Malhotra of two regions of India, Madras in the south and Punjab in the north, that appears to fly in the face of the heart–diet hypothesis.[16] In that study of over a million Indian men, the CHD death rate was found to be seven times higher in Madras than Punjab, even though in Punjab "people ate ten to twenty times more fat and smoked eight times more cigarettes."[17]

This startling comparison becomes somewhat less puzzling, however, on looking into the actual data from Malhotra's study. First, the number of cigarettes smoked is irrelevant, because the cigarette sales in both regions were so small that in Punjab (the region with more smokers), they amounted to only 14 cigarettes per person per year on average. Furthermore, the dietary comparison may also be misleading. Although Malhotra does note that fat consumption was 19 times more in Punjab, people in that region consumed only a quarter as much meat, fish, and eggs as those in Madras, with their much higher fat intake coming from milk and milk products. So, the contradiction to the heart–diet idea provided by this example is not quite as stark as Ravnskov makes it out to be.

Ravnskov notes that several African groups also provide examples that appear to stand the heart–diet idea on its head. The Masai people of Kenya are nomadic tribesmen whose diet consists entirely of milk, blood (from cattle), and meat. The Samburu people are another tribe having a similarly extreme diet. Apparently, Samburu men consume almost a gallon of milk per day, including over a half-pound of butter fat. Studies of the Masai and Samburu peoples show that despite their extremely high-fat diets, these tribesmen do not die from heart disease.[18] Even more surprising, perhaps, was the observation that the blood cholesterol levels of these people was among the lowest in the world—about half that of most Americans.

273   It is possible that tribes like the Masai are not typical of other groups, because they have a genetic predisposition toward low blood cholesterol even when their diet is very rich in animal fat. Such an explanation might lead us to predict that the Masai have even lower levels of blood cholesterol when they move to urban areas and their diet becomes more diversified. In fact, however, just the reverse is true. The urbanized Masai have cholesterol levels about 25 percent higher than those in the countryside, perhaps because of other risk factors.[19] On the other hand, the very low rates of heart disease in the Masai may, in part, have a more mundane explanation, namely their rather low life expectancy—in the mid 40s—a bit less than other Kenyans. Thus, most of the Masai simply don't live to the older age at which heart disease claims many victims!

Such "anomalies" as the Masai cannot prove that diet plays no role in coronary heart disease, but they do show that many more important factors exist. The Masai people, for example, are slender, extremely physically fit, and don't smoke. On autopsy, their coronary arteries are smooth and appear to be much wider than those of most people living in developed nations, but they do have significant amounts of atherosclerotic plaque.[20] The higher cholesterol levels found when they move to urban areas probably are associated with nondietary changes, such as a more sedentary lifestyle.

Another way we might try to determine how large a role dietary fat intake plays in the development of coronary heart disease is to look at trends over time, rather than from place to place. For example, we can see how CHD mortality changes during a period of time when animal fat becomes more or less prevalent in a nation's diet. Unfortunately, this type of data shows the same sort of conflicting results as the geographic data. We can find some cases where decreasing dietary fat occurs during a period when CHD mortality decreases and some where it increases. The problem here is the same as before, namely the many confounding variables.

One example of the ease with which we can be misled if we ignore a confounding variable is the "epidemic" of heart dis-

274 ease that was said to have occurred after World War II in the United States.[21] Since this was also a period when meat became increasingly available it might appear to support the heart–diet theory. In fact, however, during this period there was no epidemic at all—the risk of dying from heart disease was unchanged for any given age group. The real change was that people were living longer to the age when they were more likely to die from diseases of old age, such as CHD—the reverse of the situation with the Masai.

One further confounding variable in this case was that physicians in the post-war era began to catch on to the new terminology and were more likely than before to write coronary heart disease on death certificates.[22] In one single year, 1948 to 1949, the addition of the term "arteriosclerotic heart disease" to the International Classification of Disease had the effect of raising the CHD death rate by 20 percent in white males and 35 percent in white females.

One way to control for confounding variables is to do a study in which a group of coronary heart disease patients is compared to a control group in the same geographic region that is matched by age and sex. We can then see how the two groups differ in regard to a whole range of possible factors that might put them at risk of heart disease. The possible factors we could look at in such a "case-control" study is virtually unlimited. As Ravnskov notes, a large number of such studies have been conducted, and taken collectively they offer little support for the heart–diet theory. In 1998 Ravnskov examined all studies that had been done up to that time, which included 34 different groups of CHD patients and controls—a total of 150,000 indi-

Number of studies	Who ate more saturated fat?
3	CHD patients
1	Control group
30	Neither

275 viduals. A summary of the findings as presented in his book is shown at the bottom of the previous page.

In 30 out of 34 cases no statistically significant difference existed between the intake of saturated fats for the two groups according to this summary.

In fact, the average percentage of calories from saturated fats in all the studies was 16.4 percent for CHD patients, and only very slightly less (16.1 percent) for the control groups.[23] For polyunsaturated fats (which are considered "good" fats), the averages were also scarcely different: 5.9 percent for CHD patients and 5.8 percent for all the control groups.

"Good" and "Bad" Fats?

Many researchers disagree with some of Ravnskov's views on the relationship between diet and heart disease. For 20 years the Harvard School of Public Health has accumulated data on almost 300,000 Americans. Their results show (in agreement with Ravnskov) that dietary cholesterol has only a small link to cholesterol in the blood and that total fat in the diet has no link to either heart disease or health generally. But, in contrast to Ravnskov, they have found that "good" (unsaturated) fats lower the risk of heart disease, while "bad" fats increase it.[24] The bad fats include both saturated fats, and the "really bad" trans fats, which remain solid at room temperature, and are found mainly in margarine and commercially prepared baked and processed foods.[25] The trans fats have been found in seven studies to raise the ratio of "bad" to "good" cholesterol in the blood twice as much as saturated fats.[26]

"Good" and "Bad" Cholesterol?

Cholesterol is an insoluble substance that travels in your blood inside small spherical particles of fat also known as lipids. In the United States their concentration in the blood is measured in milligrams per deciliter

276 (mg/dL), where 1 mg/dL is equivalent to approximately one part in 100,000 by weight. Although the earliest studies didn't make the distinction, you are probably aware of there being several types of cholesterol—the "good" high-density lipoprotein (HDL) type, and the "bad" low-density lipoprotein (LDL) type. Typically, between 60 and 80 percent of the cholesterol in your blood is of the LDL variety. Studies that refer to cholesterol (sometimes called *total* cholesterol) levels are looking at the sum of the LDL and HDL concentrations. The two types differ not only in their density, but also in their destination. The bad guys go from the liver, the source of most of the body's cholesterol, to cells that have a need for it. The HDL goes mainly in the other direction.

What is the whole basis for calling HDL good and LDL bad? The same kind of epidemiological studies that have shown total cholesterol to be a risk factor for CHD have shown LDL to be an even stronger risk factor, and some studies have found that the ratio of HDL to total cholesterol is the strongest risk factor.[27] Obviously, if HDL is good, the higher this ratio the better. Part of the reason high values for that ratio may be good is that they are associated with many other beneficial conditions: weight loss, exercise, nonsmoking, and reduced blood pressure. But Ravnskov suggests that, like total cholesterol, this ratio is merely a marker for the real culprit, and not itself a cause of heart disease.

In addition to LDL and HDL, there are a number of other biochemicals in the blood that are related in some way to the development of coronary heart disease. Two, in particular, include triglycerides and lipoprotein (a), both of which appear to be independent risk factors for CHD, according to some studies.[28] As with LDL and HDL cholesterol, elevated triglyceride and lipoprotein (a) levels could conceivably be merely markers for CHD.

Even if saturated fat is not completely harmless as Ravnskov says, its impact on health would appear to be fairly small, according to the published literature. Furthermore, it is interesting that even those studies that show that saturated fat intake does lead to higher risk of coronary heart disease find no differ-

277 ence in mean blood cholesterol levels between those individuals who ingest little or much saturated fat, which seems to contradict the idea that elevated cholesterol leads to CHD.[29] Regarding the magnitude of the risk of death, three research groups have asked what would happen if people strictly adhered to a diet in which only 10 percent of their calories came from saturated fats. Their conclusion was that healthy non-smokers might gain perhaps 3 days to 3 months extra life expectancy.[30]

If the belief that a high-fat diet is dangerous really is just a myth, as Ravnskov claims, how did that myth gain its current status? Science journalist Gary Taubes traces the entire fascinating story in a 2001 article in *Science* magazine: "The Soft Science of Dietary Fat."[31] Apparently, the anti-dietary fat movement began in the 1960s and was promoted by a Senate committee on nutrition led by Senator George McGovern. The media jumped on the bandwagon, and scientists were asked to provide data to support the view that fat in the diet was bad: "Once politicians, the press, and the public had decided dietary fat policy, the science was left to catch up."[32]

Even by 1988 the science had not quite caught up. That was the year that the U.S. Surgeon General's Office decided to gather all the available evidence and write the definitive report on the dangers of dietary fat. The project did not come to fruition, and it was finally killed with little fanfare in 1999. The only explanation of its demise was provided by a member of the project's oversight committee who noted that "the report was initiated with a preconceived opinion of the conclusions," but the science behind those opinions was not holding up. "Clearly the thoughts of yesterday were not going to serve us very well."[33]

It would appear that Ravnskov is largely correct regarding the lack of hard evidence linking a high-cholesterol diet and heart disease. His belief regarding the complete innocence of saturated fats may be an overstatement, but their impact on CHD appears to be less than is commonly assumed. Interestingly, his

278 position on trans fats is much more cautious, noting that they have "silently infiltrated the food supply like a marauder in the night."[34]

Does High Cholesterol in the Blood Cause Heart Disease ("Myth 2")?

The claim that elevated levels of cholesterol in the blood are associated with increased risk of CHD rests on a number of studies done over the years. One of the largest of these, cleverly known as MRFIT (Multiple Risk Factor Intervention Trial), looked at more than 300,000 American men over many years. The MRFIT study examined how mortality from various causes was associated with level of blood cholesterol. One recent publication by MRFIT researchers reports on a group of 69,000 young men whose cholesterol was measured (only once) in 1973–75.[35]

Figures 9.3 and 9.4 from this publication show how the death rates from CHD and from all causes depend on cholesterol lev-

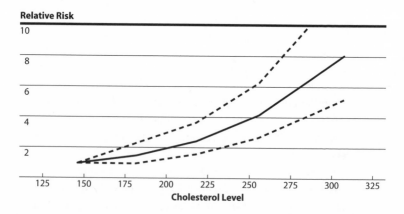

Figure 9.3 Age-adjusted relative risk of CHD death for men versus cholesterol level from the data in Stamler et al.[35] The area between top and bottom lines shows the 95 percent confidence interval.

279 els. The vertical axis in both figures is the "relative risk" of death, which has been age-adjusted. If the relative risk at some particular cholesterol level were 2.0, for example, that would mean that the age-adjusted death rate is twice that at the lowest cholesterol level. Age-adjustment of death rates is important, because cholesterol levels tend to increase with age and so does the chance of dying. Without making an age adjustment if we see a rise in relative risk with increasing cholesterol levels, we wouldn't know how much that rise was due to people with those higher cholesterol levels simply being older.

As can be seen in figure 9.3, the increase in relative risk of CHD mortality with cholesterol level is very significant and almost follows an exponential rise. The dashed curves show the 95 percent confidence intervals for each cholesterol level. As Stamler et al. note in commenting on their results, "there is a continuous, graded, strong, independent relationship of serum cholesterol level to long-term risk of CHD."[36] Not surprisingly, Ravnskov finds flaws with the MRFIT study. He first mentions a Swedish medical researcher, Lars Werko, who raises questions about the ability of the study to track so many individuals

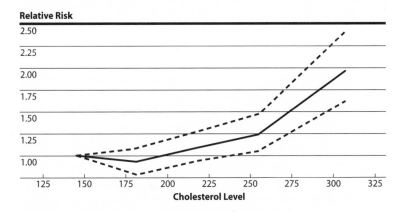

Figure 9.4 Age-adjusted relative risk of death from all causes for men from the data in Stamler et al.[35] The area between the top and bottom lines shows the 95 percent confidence interval.

280 reliably and about inconsistent values in various reports on the numbers of study participants. Most of Werko's criticisms, however, apart from a claim of falsification of data by one participating center, seem fairly minor.

Risks: Relative or Absolute?

Like many studies, MRFIT expresses risks in relative rather than absolute terms. Ravnskov believes that this practice yields a misleading (inflated) impression of the real magnitude of the risk. For example, in a 1986 MRFIT publication men in the highest cholesterol category had 4.13 times the risk of those in the lowest cholesterol category, which could be characterized as a 413 percent greater relative risk. But, in fact, when the risks were expressed in absolute terms, the risks of dying over the six-year observation period in the highest and lowest cholesterol categories were only 0.3 and 1.3 percent (which is 4.13 times greater than 0.3). The absolute 1.0 percent difference between them sounds far less ominous than 413 percent—the relative risk.

So, which way of stating risk is more relevant for real-world decision-making—absolute or relative? Let's look at another example before you decide. If you are an "average" driver, your chances of being killed in a fatal automobile accident are around 0.02 percent per year if you wear seat belts, and perhaps 0.04 percent per year if you don't. Going by absolute risk you might say that you raise your chances of being killed in a car crash each year only by a mere 0.02 percent by not wearing seat belts, so why bother—assuming you were unconcerned about getting a traffic citation. But according to a relative risk comparison your chances of dying in a car crash are twice (200 percent) as great if you don't wear seat belts. You can decide for yourself which way of stating the risk is the more relevant one for seat belts. In the case of coronary heart disease, a difference of a "mere" 1 percent in the death rate over a six-year period translates into a difference in life expectancy measured in at least four years. Do you think Ravnskov's preference for absolute risk is justified?[37]

281   Ravnskov is also critical of the MRFIT study because he claims that the increase in risk is mainly confined to the highest cholesterol levels. He notes that "in most studies, the increased risk is present only above a level of cholesterol that includes just a small fraction of the population." In a strictly technical sense Ravnskov is right on this point. Most published studies involve far fewer participants than MRFIT and therefore would not have the statistical power to show significant increases in relative risk until much higher cholesterol levels, where the risks are greatest and can most easily be seen. But we cannot then conclude that we should doubt the MRFIT results just because most studies show no increased relative risk at moderate cholesterol levels, if they didn't have the needed statistical power to do so! Moreover, many large studies besides MRFIT *do* show a gradual increase in risk with rising cholesterol levels, including the PROCAM study, the Chicago Heart Association Study, the People's Gas Company Study, and the Framingham study.[38]

Framingham is a small town in Massachusetts, the site of a study begun in the 1950s to investigate the risk factors for coronary heart disease. Unlike the MRFIT study, Framingham included both men and women as participants, and it monitored participants' cholesterol levels over an extended time, not just once. The results of Framingham are also often cited in support of the conventional view that high cholesterol leads to higher incidence of CHD, but the actual picture is more complex.

For men in the Framingham Study who were under age 48 it was found that there was a 5 percent increase in the death rate for each 10-mg/dL increase in their cholesterol level. Thus, if your cholesterol level were 240 you would, in principle, have a 20 percent higher chance of dying per year compared to someone whose level was 200. (As we shall see, however, it does not necessarily follow that your chance of dying would drop by 20 percent if your cholesterol level dropped spontaneously from 240 to 200.)

What about Older Men? Framingham men over 48 years old having high cholesterol showed no statistically significant risk

282 of mortality either from cardiovascular disease or from all causes.[39] Other studies besides Framingham also support the claim that if older men are at increased risk of death due to elevated cholesterol, the extra risk is not high. For example, Krumholz et al. found that for nearly a thousand subjects over 70 years of age high cholesterol did not elevate their risk of death over a four-year period.[40] Specifically, they found that the highest cholesterol group had a relative risk (compared to the lowest cholesterol group) of 0.99 with a 95 percent confidence interval of risk: 0.56 to 2.69. What that really means is that these elderly high-cholesterol men might have anywhere from 56 to 269 percent the risk of dying as the lowest cholesterol elderly men, which represents a rather large uncertainty range centered on no increase in risk. With such a large uncertainty range we cannot say confidently that there is no elevated risk, only that it cannot be measured within certain limits. In particular, if there is a risk due to high cholesterol in men over 48, it is less than half that found in younger men.

Usually studies that look at the *incidence* of coronary heart disease rather than mortality obtain more precise results, because there are more cases of CHD than fatalities in a given time period. For example, according to data from the Framingham Study, men aged 65 and older in the highest cholesterol decile had a relative risk for CHD of 1.8 compared to those with cholesterol levels below 200 mg/dL.[41] The 95 percent confidence interval was 1.3 to 2.5 in this case. This result was supported by an Australian study in which high cholesterol men over 60 were found to have a relative risk of 1.24 (95 percent confidence interval: 1.06 to 1.46). Interestingly, Ravnskov cites this same Australian study to make the *contrary* claim that cholesterol is not a risk factor for the elderly, because there was no elevated risk found for men above 74. (Obviously, if the study statistics were just barely sufficient to see an elevated relative risk for men over 60, they would not be likely to have sufficient power to see anything for men over 74.) Thus, Ravnskov's claim notwithstanding, high cholesterol is a risk factor for older men,

283 even if the elevated risk with rising cholesterol levels is only perhaps half that for younger men.

What about Women? Ravnskov claims that "most studies have found that high cholesterol is not a risk factor for the female sex."[42] Let's take a look at the Framingham data for women to see how well this claim stands up. The Framingham data that looked at the women's incidence of coronary heart disease (rather than death) found a very statistically significant increased risk for elevated cholesterol levels for women aged 48, though it was not as pronounced as that for men (figure 9.5). The Framingham data also suggest that women in their 50s are at increased risk of death at *both* very high and very low cholesterol levels (figure 9.6). It greatly distorts the Framingham data to claim that cholesterol is not a risk factor for women. However, given the smaller elevation of CHD risk seen for women compared to men in the Framingham data, Ravnskov is *technically* correct that most studies that are often statistically

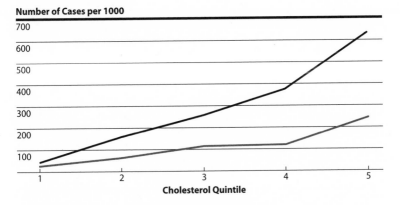

Number of Cases per 1000

Cholesterol Quintile

Figure 9.5 Eight-year incidence of CHD for men (upper curve) and women (lower curve), at age 48 years, according to the data in W. B. Kannel and D. L. McGee, Composite scoring—methods and predictive validity: insights from the Framingham Study. *Health Services Research* 22(4): 499–535 (1987).

284   less powerful do not show any increased risk for women due to high cholesterol.

Table 9.1 shows the CHD risks for various ages and both sexes, according to the National Institutes of Health.[43] Note that while the CHD risk is greater in the old than in the young, the rate of increase in risk with rising cholesterol levels is greater in the young.

An "Alarming" Finding? Ravnskov believes that he has uncovered an "alarming" finding in the Framingham study that directly contradicts the heart–diet idea, even though most people regard the study as strongly supporting it. The Framingham study reports the following rather surprising finding: "For each 1-mg/dL per year drop in serum cholesterol values over the 14-

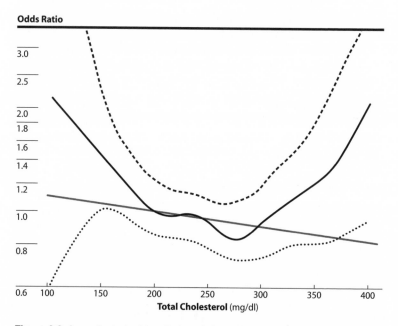

Figure 9.6 Age-adjusted odds ratio (or relative risk) of mortality from all causes for women versus their cholesterol level. From Emond and Zareba,[47] printed with permission from the *Journal of Women's Health*. Upper and lower curves show the limits of the 95 percent confidence interval.

285 Table 9.1

Ten-Year %Risk of CHD by Cholesterol Level for Men and Women

Age	Men			Women		
	<200	*200–39*	*240+*	*<200*	*200–39*	*240+*
40	3	5	12	1	2	5
50	8	10	15	2	4	8
60	16	15	21	5	8	11
70	18	22	28	5	7	13
80	14	23	29	14	16	17

year period of cholesterol measurement, there is an 11 percent increase in both the overall death rate . . . and the CVD [cardiovascular disease] death rate."[44] Ravnskov wonders how the study could possibly be used to support the heart–diet idea if those whose cholesterol had *decreased* during the study ran a greater risk of dying than those whose cholesterol had *increased*.

There are at least two possible explanations for this strange Framingham result. One possibility is that those participants in the Framingham Study whose cholesterol decreased during the 30 years of the study had much higher cholesterol levels to begin with than those whose cholesterol levels didn't decrease. In that case, the former group would have had higher average cholesterol levels over the 30 years and would be expected to show greater mortality risk. This possibility seems unlikely for two reasons: (1) the Framingham authors didn't mention it, even though they would have had every reason to look for such a mitigating factor, and (2) during most of the 30-year period when the study was done there were relatively few patients on cholesterol-lowering drugs.

A second possible reason why Framingham participants whose cholesterol dropped had a higher death rate is the one favored by the study authors. They suggest that spontaneous drops in cholesterol levels can be a marker for the onset of cer-

286 tain fatal illnesses. The authors go on to suggest that this same effect might explain why the elderly with high cholesterol show less increased risk than the young. Their lower elevated risk could be due to two offsetting trends: increased risk due to elevated cholesterol levels, and decreased risk among those whose levels do *not* drop spontaneously. This explanation could be correct, because the Framingham data showed substantially increased odds of cancer death if cholesterol drops by a large amount over any 4- to 6-year period.[45] In addition, this explanation is credible because other risk factors for mortality, such as excess weight or iron in the blood, also become risk factors when one experiences a spontaneous unexplained *decrease*. On the other hand, the study found that falling cholesterol levels also were associated with increased mortality for cardiovascular disease. The explanation given by the authors of a possible "lipoprotein metabolic dysfunction" seems plausible if somewhat ad hoc.

Ravnskov draws a different conclusion from the association found between dropping cholesterol levels and excess mortality, which he views as an outright contradiction to the idea that high cholesterol is dangerous. He suggests that if dropping cholesterol levels can be considered a marker (rather than a cause) for some serious underlying condition, the same could be true of elevated cholesterol levels themselves. In other words, if we are so certain that dropping cholesterol levels cannot really be a cause of disease, what makes us so sure that high levels to begin with are a real cause of CHD? Cholesterol, in Ravnskov's view is just an "innocent bystander" present at the scene of the crime.

Guilt by Association? Proponents of the heart–diet idea need to do two things to show that high cholesterol is a cause of coronary heart disease and not merely a risk factor. First, they must establish that high cholesterol is a risk factor for CHD that is *independent* of other risk factors, such as age, smoking, obesity, high blood pressure, and stress. This independence, in fact, *has* been shown in studies that look at correlations between all

287 the variables that affect mortality. The Framingham study, for example, found that when a multivariate analysis was used to adjust for these other risk factors, the risk due to cholesterol still remained, although the actual risk dropped by about 20 percent.

What Are Your Chances?

Do you want to find out what your chances are of getting coronary heart disease in the next 10 years according to the heart–diet theory? The Framingham Heart Study has developed a score sheet that can be found at www.nhlbi.nih.gov/about/framingham/riskabs.htm.

The score sheet calculates your chances based on eight different factors.[46] According to the score sheet for men, each of the following increases your risk by the same amount: being a smoker, having diabetes, being ten years older, having 20 mm higher systolic blood pressure (the higher of the two readings), having a cholesterol reading higher by 70 mg/dL, or an HDL reading *lower* by 15 mg/dL. For women, diabetes is given twice as much weight as for men, and cholesterol is given a little less weight. Also, the weight given to age for women changes drastically above 55 years, but at all ages the average risk of CHD for women is a bit less than half that for men.

Strangely, the Framingham score sheet for women indicates higher risk only for elevated cholesterol levels. It is therefore inconsistent with the study result that for women aged 56–70 elevated cholesterol poses just as much risk as very low levels, and for women above 70 *low* cholesterol levels appear to be the primary risk factor.[47] The score sheet is also inconsistent with their findings showing that the increase in risk of CHD with rising cholesterol levels is much less for elderly men than younger men.

Another more serious problem with proving that CHD results from high cholesterol is that a world of difference exists between "risk factors" and causes. Risk factors can be totally innocent bystanders that merely are associated in some way

288 with a condition, such as CHD, without having any effect whatsoever. As Ravnskov notes, owning a television set can be considered a risk factor for coronary heart disease, because the per capita ownership of TVs correlates with a nation's prosperity, and therefore with the incidence of CHD. (That's because more prosperous nations tend to have more industrial pollution, more cars, more obesity, and more people living to old age, all of which increase the incidence of CHD.) Even though television ownership is a risk factor for CHD, it would be foolish for you to try to lower your risk by throwing out your TV, although that might actually have some other benefits.[48]

Since none of the studies showing that high cholesterol is a risk factor for CHD show it to be a cause, it is not yet obvious that Ravnskov is mistaken in his belief that it is unwise to combat CHD by lowering your cholesterol. The separate but closely related issue of the value of the cholesterol-lowering drugs will be discussed later, and we will see that those results do shed some light on Ravnskov's claim.

Ravnskov has certainly not proven that elevated levels of LDL, total cholesterol, triglycerides, and lipoproteins are innocent bystanders—risk factors that are merely associated with the development of CHD. But, he doesn't have to! It is enough for him to point out that heart–diet proponents have not made their case that these risk factors for CHD are causes. Proving causation is extremely difficult for any disease—even the link between smoking and lung cancer took years to establish. Still, the diet → CHD link would seem to be far less well established than either HIV → AIDS or smoking → lung cancer.

Is High Cholesterol Caused by Bad Diet ("Myth 3")?
Does It Block Arteries ("Myth 4")?

Ravnskov believes that high cholesterol in the blood has little relation to your intake of saturated fats or cholesterol. He also believes that arteries are not blocked as a result of high cholesterol levels in the blood. Let's first consider the evidence concerning the linkage between diet and cholesterol levels in the

289 blood. Results from various published studies do substantiate Ravnskov's claim that there is only a weak connection between the diet and cholesterol levels in a group of individuals. These include a study of over 2000 people in Michigan, three smaller American studies, a Finnish study, and a large Israeli study of 10,000 men. The Israeli study, in fact, found that the correlation between animal fat intake and the level of cholesterol in the blood was exactly 0.0—no correlation whatsoever.

Are Heart–Diet Proponents Biased?

Ravnskov believes that proponents of the heart–diet theory disregard much contrary evidence and, even worse, misreport the results of their own studies to conform to the conventional theory. Among the numerous examples of such reporting bias, he cites a paper by Ascherio et al. on the relation between dietary fat and risk of CHD.[49] In the conclusions section of the paper the authors note that "saturated fat and cholesterol intake affect the risk of coronary heart disease as predicted by their effects on blood cholesterol concentration." The data presented in the body of the paper, however, seem to be at odds with this assertion. In table 1 of their paper the authors show that the mean blood cholesterol levels for each value of saturated fat intake are virtually identical. Upon contacting the lead author of this paper to inquire about this apparent contradiction I was told that "people with (genetically determined) high cholesterol may have reduced their saturated fat intake to lower their cholesterol and this could offset the expected positive association."[50] Such an explanation, while theoretically possible, would seem to require a highly coincidental offsetting effect. In this case and some others Ravnskov's claim of reporting bias would seem to have validity.

Ravnskov quotes a pair of researchers connected with the Framingham Study (William Kannel and Tavia Gordon) who summarized their findings that a participant's diet is not the source of the differences found in cholesterol levels between individuals in the Framingham Study. He also notes that for un-

290 known reasons these particular results were unpublished. My contact with one of these researchers, William Kannel, has confirmed the essential correctness of Rasnkov's quote. As Kannel explains, "What we were pointing out was that at the time the Framingham Study population was universally exposed to a high fat diet yet there was a wide distribution in the cholesterol values. Hence we believed that there must be a variation in the ability to cope with this noxious exposure [to cholesterol], or that there was some additional factor that was involved."[51]

The lack of a strong relation between diet and cholesterol from one individual to another does *not* imply that you cannot change your cholesterol level by switching to a low-fat diet. When such interventions are made, however, the results are usually fairly modest, at least when diet alone is changed. You might find a reduction of perhaps 10 percent in cholesterol as a result of an extreme reduction in animal fat, while for a more palatable diet you are likely to experience a reduction of only a few percent.[52] One intervention trial (MRFIT) did report a 7 percent decrease in blood cholesterol levels over seven years—after reductions in its dietary intake of cholesterol by 40 percent and saturated fats by 25 percent.[53] But in a control group whose diet had scarcely changed the levels drop by 5 percent. Perhaps the most conservative way to state the MRFIT result would be a drop of between 2 and 7 percent owing to dietary changes alone, since it is not clear how much the control group may have lowered its cholesterol levels due to other lifestyle changes.

Some readers may have experienced greater cholesterol reductions than a few percent without drugs, but in that case you probably made other lifestyle changes as well, including weight reduction, smoking cessation, stress reduction, and/or establishing an exercise program. It probably should not be too surprising that we can have only a modest effect on cholesterol levels through diet alone, because the body compensates by producing more cholesterol when we take in less, and vice versa. Ravnskov would appear to be correct with regard to his "myth 3" that diet (independent of other lifestyle factors) usually has only a modest relation to cholesterol levels. On the other hand, it is also possible

291   that a low animal fat diet is capable of lowering cholesterol, but in practice only modest gains occur, possibly because most people in Western societies are incapable of adhering to a radical departure from their typical dietary patterns. But even if Ravnskov were mistaken here, it wouldn't matter, *if* he is correct in his claim that high levels of cholesterol in the blood do not cause arteries to become blocked (atherosclerosis)—"myth 4."

Does High Cholesterol Cause Atherosclerosis? Of all the beliefs about cholesterol, the one that appears to be most widely accepted, and which Ravnskov disputes most strongly, is that high levels of cholesterol in the blood are responsible for atherosclerosis. Radioactive tracer studies have shown that cholesterol that winds up in the walls of blocked arteries was carried there in the blood, rather than being synthesized in the arterial wall. But that observation in no way shows that elevated blood cholesterol *caused* the arteries to become blocked.

One way to test whether high cholesterol levels are responsible for atherosclerosis is to see how the two variables are correlated in a group of people. If high cholesterol causes atherosclerosis, it might seem that the two variables should be highly correlated. The data in figure 9.7 were obtained by autopsying

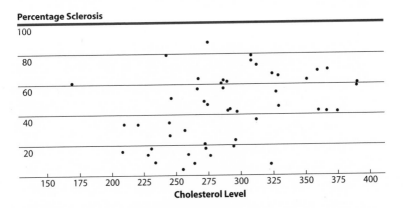

Figure 9.7 Extent of atherosclerosis versus cholesterol level. From Ravnskov,[1] printed with permission.

292 a sample of 50 middle-aged men. As can be seen there is very little correlation between the degree of sclerosis and the cholesterol level measured at death, with the correlation coefficient between the two being only 0.29.

Ravnskov notes that the degree of correlation would be even less if we excluded those individuals with the rare genetic disease of familial hypercholesterolemia (FH), which impairs one's ability to metabolize cholesterol properly. FH predisposes people to have both heart disease and very high cholesterol levels, but through a mechanism that may have no relation to how coronary heart disease develops in the 99 percent of people who are free of this disorder. If we assume that the data points above a cholesterol level of 350 in figure 9.7 represent people with FH and we exclude those points, we find essentially zero correlation between cholesterol and arterial blockage—the points form a random pattern.

Such an arbitrary post hoc exclusion of data is questionable, however, because it seems unlikely that so many individuals in a sample of 50 would have the condition of FH. Nevertheless, it is important to appreciate that spurious correlations can arise in such plots, depending on how the sample is selected. For example, it would be unwise to include a sample of people who died at a wide range of ages, because the elderly tend to have both more atherosclerosis and higher cholesterol levels. This kind of selection again would give a clump of data points that lie in the upper right quadrant of the plot.

Ravnskov cites a number of other studies that found no correlation between cholesterol level and degree of atherosclerosis. But he also notes that still other studies have found some positive correlation, including the Framingham Heart Study, which reported a correlation coefficient of 0.36. Ravnskov suggests that even that modest correlation may be spurious, however, because Framingham selected only 14 percent of individuals in their study to be autopsied and did not say anything about their selection criteria, raising the possibility that the group was a biased sample. As noted above, a spurious correlation can be created by including a disproportionate sample of

293 people with the genetic disorder FH, or by including people who died at very different ages.

Given the small degree of correlation seen in these studies overall—Framingham was the largest at 0.36—you might think that the hypothesis that atherosclerosis is caused by high blood cholesterol is on shaky ground, as Ravnskov claims. However, you would be mistaken! The primary reason for a small degree of correlation between blood cholesterol levels and degree of artherosclerosis is that the latter depends significantly on *time*—presumably the time since some initiating event, perhaps an injury to the arterial wall. In other words, an individual having any particular cholesterol level could have very different amounts of clogging of the coronary arteries, depending on when he or she first became injured. Person A with very high cholesterol, whose arteries were first injured last week, would have much less of a developed lesion than person B with low-cholesterol, whose arteries were first injured ten years ago. By averaging over many individuals whose arterial injury occurred at different times, you smear out and therefore greatly dilute the contribution due to differences in their cholesterol levels. Contrary to Ravnskov's claim, the lack of a large correlation between cholesterol levels and degree of atherosclerosis is not evidence against a causal connection between them.

Another Japanese Paradox?

Earlier we considered the interesting case of the Japanese, who have among the lowest incidence of CHD on the planet, unless they migrate to the United States. Studies have been done on the levels of blood cholesterol and the degree of atherosclerosis of Japanese (who died in Japan). As expected, the Japanese were found to have much lower blood cholesterol levels than Americans, and also much less sclerosis of the coronary arteries. But, the "shocking" fact was that compared to Americans, they had essentially the same degree of sclerosis in the aorta, the main artery in the body, and were even more sclerotic than Americans in their brain arteries.[54] Ravnskov wonders why, if high levels of cholesterol really

294 are the cause of sclerosis, Japanese with their low cholesterol levels could possibly have more sclerotic brain arteries than Americans and the same degree of aortic sclerosis? This paradox suggests at most, however, that cholesterol levels are not the *only* determinant for the development of atherosclerosis. Perhaps differences in arterial diameter and extent of branching, and blood pressure in the respective arteries might account for the differences found. Thus, the "paradox," while interesting, doesn't disprove the conventional theory.

What Can We Learn from Animal Experiments? As already noted, only modest correlations are seen between natural blood cholesterol levels and the degree of atherosclerosis. Much information about the development of atherosclerosis and its possible relation to cholesterol levels might be learned if it were possible to raise our cholesterol level dramatically, but such experiments obviously cannot ethically be done on humans. Ethical or not, society does sanction animal experiments, in which various creatures have been fed tremendous amounts of cholesterol.

In nature, many mammals are vegetarians and never eat cholesterol as part of their ordinary diet. Strangely, while a form of athereosclerosis does occur naturally in some vegetarian birds, such as pigeons, it is not seen in beasts of prey that do ingest cholesterol. Ravnskov believes that this observation is difficult to reconcile with the conventional heart–diet idea. When researchers try to manipulate animal cholesterol levels in controlled experiments they often use vegetarians, such as rabbits, to study the effects of elevated cholesterol in creating atherosclerosis.

Animal vegetarians have the property that their bodies cannot digest cholesterol. When they are force fed the stuff their blood cholesterol levels skyrocket to perhaps 20 times the highest values seen in humans. For such animals, "every organ soaks up the cholesterol like a sponge soaks up water."[55] Interestingly, while the elevated cholesterol causes the rabbits to die eventually—mainly from loss of appetite and emacia-

295 tion—it does not lead to arterial changes that resemble those seen in human atherosclerosis.

When baboons are used in such cholesterol-feeding experiments, the result is much the same—no arterial lesions that resemble human atherosclerosis. On the other hand, pigs and rhesus monkeys fed a high cholesterol diet have developed arterial changes that do resemble those seen in humans—in support of the heart–diet idea.[56] Yet Ravnskov wonders: "How do we know whether man reacts like a rhesus monkey or like a baboon or in some other way?"[57] In other words, he suggests that it would be circular reasoning to argue that rhesus monkeys and pigs must be the appropriate human model, simply because it is in those creatures that scientists have induced atherosclerosis by manipulating cholesterol levels. Actually, it is not circular reasoning. Only certain animals do develop arterial lesions that strongly resemble those seen in humans, and it is in those specific cases that the lesions develop when the animal's cholesterol levels were increased. This observation is not conclusive evidence that high cholesterol really is the cause of atherosclerosis in humans, but it is certainly supporting evidence. Ravnskov's view that conflicting results from animal experiments make them entirely irrelevant to cholesterol's role in the development of atherosclerosis in humans is unwarranted.

Why Is There No "Exposure–Response" Relation? One reason Ravnskov offers for being especially skeptical that high cholesterol causes atherosclerosis in humans is the lack of an "exposure–response" relation between the two variables. He takes the exposure–response relation to mean that whenever one factor (here high cholesterol) increases, then the condition it supposedly causes (here atherosclerosis) should increase. But, as Ravnskov notes, none of the studies show such a relation: "coronary atherosclerosis gets worse just as fast or faster when cholesterol goes down as when it goes up, the opposite of exposure–response." This absence of an exposure–response relation alone "should have led scientists to question the whole diet–heart idea."[58] Is he right?

296   In applying the exposure–response idea to cholesterol and atherosclerosis, let us assume that the rate at which plaque forms in the coronary arteries is proportional to the cholesterol level in the blood, which may vary over time. The rate, of course, also may depend on other factors, including blood pressure and the presence of an initial injury to the arterial wall. For simplicity, let's ignore here any factors that might counter the development of atherosclerosis. The total amount of sclerosis will then be proportional to the *average* cholesterol level multiplied by the total time plaque has been accumulating. Now, suppose cholesterol levels go down in a group of patients. In that case the *rate* of accumulation of plaque would be reduced, but plaque would still continue to accumulate at a reduced rate. It certainly does *not* follow from the exposure–response relation—as Ravnskov claims—that atherosclerosis should decrease in patients whose cholesterol goes down, only that its rate of increase should slow. Ravnskov seems to be confusing a change in the rate of accumulation with a change in the total amount of accumulation.

Although Ravnskov appears to be mistaken regarding his exposure–response argument being the "nail in the coffin" against a causal link between elevated blood cholesterol and the development of atherosclerosis, some of his other arguments do raise a bit of doubt on the matter. If cholesterol is not the whole story (or conceivably even an innocent bystander), we are tempted to ask who is the guilty party? During the past decade some exciting research has suggested that atherosclerosis might be caused, at least in part, by an infection. In fact, intact microbes have been found in arterial plaque, and those same microbes have initiated and accelerated the development of atherosclerosis in animals.[59]

Will Lowering Cholesterol through Drugs Lower Your CHD Risk?

At one time cholesterol-lowering drugs were prescribed only for persons with severely elevated cholesterol levels. But in recent years a massive campaign has been underway in the United States and other developed nations to use drugs to lower

297   cholesterol levels for otherwise healthy people who have moderately elevated cholesterol, but who do not have coronary heart disease. The current generation of these drugs, known as the statins, began to gain widespread use during the late 1980s. The statins are known by trade (generic) names, and include Lescol (fluvastatin), Lipitor (atorvastatin), Mevacor (lovastatin), Pravochol (pravastatin), and Zocor (simvastatin). Current U.S. sales of statins are over $10 billion, and worldwide sales are expected to top $25 billion over the course of the next few years. These figures dwarf those of such highly publicized medications as Viagra or Prozac.

Just for the sake of argument, let's imagine that CHD really is caused by an infectious organism as some research suggests, and further that it might be prevented by a vaccine—I know it's unlikely, but humor me! Sales of such a one-time vaccine would be miniscule compared to the continual flow of sales to supply the daily doses of the current generation of cholesterol-lowering drugs. Is your faith in the humanity of drug company executives so great that you would imagine them eager to fund trials of potential vaccines that would jeopardize their current lucrative market? If so, I have a bridge I'd like to offer for sale at a great price.

Drugs to lower cholesterol have been sought since the 1960s. Some of the substances tried in those early trials included nicotinic acid, clofibrate, thyroid hormone, and estrogen (the female sex hormone). Most of the early drug trials had very modest results and, in some cases, it was found that the drug was causing rather than preventing heart attacks, as in the case of estrogen and thyroid hormone. Ravnskov discusses a number of these early drug trials, and he convincingly shows how their benefits were hyped, while their risks were minimized. In a meta-analysis of 26 controlled trials conducted before 1990, he shows, in fact, that there was no overall difference in outcome between those people on the drug or those in the control groups, whether outcomes are defined in terms of risks of heart attacks or deaths.[60]

But what about the risks of the statins, the current generation of cholesterol-lowering drugs? These drugs have received

298 a great deal of attention in the medical community, because they lower cholesterol by 30 percent or more and supposedly lack the side effects present with some of the older drugs. Equally impressive, they show similar reductions in the number of heart attacks both in CHD patients and healthy individuals, including those having average cholesterol levels. The statins, therefore would seem to pose an extreme challenge to Ravnskov's claims. If cholesterol is really an innocent bystander, how is it possible that drugs that reduce cholesterol levels by 30 percent also "coincidentally" reduce heart attacks by 30 percent? How does Ravnskov meet this extreme challenge to his ideas? After discussing the results of six trials of statin drugs that were found to lower both cholesterol and CHD, he tries to show that the trials actually prove exactly the opposite of what they seem to prove, i.e., they prove that cholesterol really doesn't matter!

Ranskov offers six arguments for this highly counterintuitive belief. The italicized text following each argument addresses the major flaws in each of his claims.

1. *Women.* The statins were almost as effective for women as they were for men, which contradicts the finding that high cholesterol is not a risk factor for women.

As noted previously, the assertion that high cholesterol is not a risk factor for women is questionable. Some studies have shown this to be true, while others have not. In the Framingham Study, high cholesterol is found to be a risk factor for CHD in women, but the rise of risk is perhaps only half as steep as for men. In four statin trials the overall reduction in relative risk of heart attack was 29 percent for women and 31 percent for men. The respective 95 percent confidence intervals were 13 to 42 percent and 26 to 35 percent. These uncertainty ranges are sufficiently broad that they are consistent with the risk reduction for women being about half that for men.

2. *The elderly.* Older persons were found to benefit from the statins as much as the young, which contradicts the finding that they are not at risk from high cholesterol.

299   *As shown earlier, this claim is false for the same reason that the first one is: the smaller elevated risk for the elderly only shows up in studies that have the necessary statistical power. However, as for women, the rise in risk with increasing cholesterol for those over 50 is only half as steep for the elderly. In four statin trials the overall reduction in relative risk of heart attack was 32 percent for those over 64 and 31 percent for those under 65. The respective 95 percent confidence intervals were 23 to 39 percent and 24 to 36 percent. Assuming that the under 65 group included many persons between 50 and 65, these results could easily be consistent with a factor of two greater risk for high-cholesterol for those under 50.*

3. Strokes. The statins reduce strokes as well as heart attacks, but high cholesterol is a weak risk factor for strokes or none at all.

The evidence linking high cholesterol and risk of strokes is uncertain. One large meta-analysis found no association between high cholesterol and increased risk of stroke.[61] But there are various subtypes of strokes and biases in these studies, such as an acute decline in cholesterol following a major stroke. In those case-control studies that included minor strokes, which do not suffer from these biases, high cholesterol was found to be a risk factor for stroke.[62]

4. CHD patients. Patients who had already suffered a heart attack benefited from the statin drugs, even though for such CHD patients high cholesterol has been shown to be a weak risk factor or none at all.

The basis for Ravnskov's claim that heart attack patients are not at risk from high cholesterol is his observation that there are some studies that show an elevated risk and others that do not. He cites 7 studies showing no risk and 13 that do show a risk, including 3 where the risk is "strong and graded." As with women and the elderly, we should not be surprised if some studies that lack the statistical power fail to show an elevated risk. There is no contradiction between positive and negative results if the negative results are negative merely because of their broader confidence intervals. It is not true that the overall result from a collection of studies should be considered as being inconclusive just because some gave negative results.

300 **5. *Protection for all.*** In the statin trials protection against CHD occurred both in those with high cholesterol and in those with low cholesterol, even though most studies show that those with normal cholesterol are not at elevated risk of CHD.

As already noted, this claim is false. There is a strong and continuous relation between cholesterol level and CHD risk, as seen in figures 9.3–9.5. It is technically true (but irrelevant) that "most" studies (including those that lack the statistical power) may not show the increased risk until very high cholesterol levels are reached.

6. *No "exposure–response."* No association was found in the statin trials between the degree of cholesterol-lowering and the reduction in risk. The risk of a heart attack was just as great for patients whose cholesterol was lowered a great deal as for those who experienced only a small degree of lowering. This is not possible if cholesterol is really the cause of CHD.

It is true that the statin trials showed no risk reduction that depended on the degree of cholesterol lowering, and in that sense they showed no "exposure–response." But this observation lacks the supreme importance that Ravnskov seems to think. If you lower your cholesterol then you do lower your risk, at least according to the conventional theory. But the trials don't measure how much people lower their risk, only what their actual risk is after so many years. The risk depends not on the cholesterol level reduction, but on the time-averaged cholesterol level over some long period of time—perhaps the time since some initial injury occurring in an arterial wall. This time-averaged cholesterol level bears no relation to the amount the level was reduced during the trial, and so we would not expect the latter to be related to a person's risk of heart attack at the end of the trial.

Ravnskov has tried through the preceding six arguments to make the difficult case that the effectiveness of the statins in reducing CHD has nothing to do with their effect on reducing cholesterol, but, as shown above, his arguments are flawed. On the other hand, statins are powerful drugs that could, in principle, have an effect on atherosclerosis independent of their cholesterol-lowering effect, as some animal experiments have sug-

301 gested.[63] And there is also recent evidence that they work for humans in part by limiting inflammation in the arterial wall, independent of their cholesterol-lowering effect.[64] But Ravnskov simply hasn't demonstrated that their cholesterol-lowering property is irrelevant to the development of CHD.

Should You Lower Your Cholesterol through Drugs?

Even if Ravnskov were right that the statins lower your risk of CHD in spite of (not because of) their effect on cholesterol, why shouldn't we all take them anyway if they lower the incidence of coronary heart disease? The only possible reason would be the cost, both economic and healthwise. Statins are not cheap. Compared to a daily dose of baby aspirin they cost about twenty times as much. Daily aspirin therapy has been demonstrated to have a relative risk reduction of 28 percent for coronary heart disease in men over age 50—very close to that for the statin drugs.[65] There is a trade-off with aspirin in a slightly increased risk of stroke and gastrointestinal bleeding, but the trade-off becomes increasingly favorable the higher your risk of CHD as measured by the usual risk factors.[66]

If you can easily afford the economic cost of the statins, your main concern would probably be the possibility unpleasant side effects. Interestingly, one possibly pleasant side effect of a heart medication was the basis for the accidental discovery of the drug Viagra. Erections occurring at embarrassing times are probably among the least serious concerns about side effects, however. While a 1990 meta-analysis by Muldoon did show that older cholesterol-lowering drugs were associated with some nasty side effects, a more recent analysis by the same author showed no such effects for the statins.[67] These drugs also have been shown in some of the clinical trials to reduce not only the number of deaths due to coronary heart disease, but also the number of deaths due to *all* causes. A meta-analysis of five trials gave a reduction in risk of death from all causes of 21 percent, with a 95 percent confidence interval of 14 to 28 percent.[68] (The risk reduction

302 was, of course, over the time of the trials, since no drug can lower your risk of dying permanently!)

Long-term cancer risk is a particular concern of many people. So, let's take a look at some of the evidence on each side of the ledger regarding this question. One study cited by Ravnskov did find an elevated risk of breast cancer for those taking statins, but the excess cancers in that study can be probably attributed to an abnormally low number of cancers in the control group.[69] Another contrary study showed that the statins may actually be protective against this particular form of cancer.[70] But perhaps the best evidence on the question of cancer and statins is provided by a recent meta-analysis of trials for five statin drugs, which showed no elevated risk for cancer of any type for those taking the drugs over a five-year period.[71] Five years, of course, is not a very long time, considering that most cancers have a latency period of 20 years or more, so the statin trials with human subjects won't be able to answer the question of the possible long-term carcinogenicity of these drugs for some time to come.

What is more worrisome, however, is that animal experiments show that the statins have caused cancer in rodents.[72] Yet the question of extrapolating carcinogenicity from rodents to humans is a controversial one. Some substances that cause cancer in rodents also cause cancer in humans, while others may not. On the other hand, almost all human carcinogens have been found to cause cancer in mice or rats.[73] It all depends on the dose. All sorts of substances have been found to cause cancer in rats if the dose is high enough. But—now here's the shocker—*the doses at which some of the statins have been found to cause cancer in rodents are close to the equivalent doses now recommended for human use.*[74] The same cannot be said for any other class of drugs intended for long-term usage. Even more worrisome, when researchers looked at the relative carcinogenicity of 80 possible chemicals to which humans are exposed, one particular cholesterol-lowering drug, clofibrate, was ranked second on the list.[75]

In view of the preceding facts, you may be wondering how is it that such cholesterol-lowering drugs ever received FDA ap-

303 proval? To answer that question, researcher Thomas Newman obtained (under the Freedom of Information Act) the minutes of meetings of the FDA Advisory Committee that evaluated two of these drugs, gemfibrozil and lovastatin.[76] In one case it appears that the animal data on carcinogenicity was presented to the committee using an older method for evaluating the equivalent human dose, based on human versus animal body mass. This older method made it appear as if the drug was about a hundred times less carcinogenic than if the newer method based on blood concentration had been used.

For the other drug evaluated by the committee, concern over its carcinogenicity was expressed. In fact, the minutes of the meeting state that only "three of the nine members [of the advisory committee] believed that the potential benefit of using Gemfibrozil for prevention of coronary heart disease outweighed the potential risk associated with such use."[77] Despite the negative recommendation from a majority of the advisory committee, the drug was subsequently approved by the FDA, and is now in widespread use.[78]

Few things frighten people more than the prospect of getting cancer from a drug they are taking to *prevent* another disease. Admittedly, even the researchers who raise the question of the possible carcinogenicity of the statins don't claim that rodent studies prove them to be carcinogenic in humans, only that they could be. Nor do these researchers say that patients at high risk of CHD should be wary of taking statins, because the benefits almost certainly outweigh the potential risks. Still, the facts about the FDA approval process for these drugs raise all sorts of disturbing questions.

Conclusions

Which of Ravnskov's "myths" about cholesterol seem to be best and least supported by the evidence? He has presented some convincing data that dietary cholesterol and fat do not, by themselves, have a very significant effect on blood cholesterol

304 and on the risk of CHD. But he seems to have not made the case on his claims that cholesterol level is not a significant risk factor for CHD and that it is a totally innocent bystander in the development of the disease. Nor has he made a convincing case that the statins reduce CHD by some means other than their impact on cholesterol level. Ravnskov did, however, help me rationalize my previously uninformed decision to stop taking my cholesterol-reducing medication, since I am not at particularly high risk of coronary heart disease, being a reasonably fit nonsmoker over 60 with cholesterol around 240. More importantly, I do not wish to be a guinea pig for perhaps the greatest medical experiment in history: a test of the long-term carcinogenic risks of a drug found to cause cancer in rodents. If I'm wrong, I could live to regret it, because while a fatal heart attack may be a good way to go compared to cancer, debilitating congestive heart failure—sometimes the result of CHD—is not.

I'm inclined to rate the idea that high cholesterol is not worth worrying about at 2 flakes. This rating should not be taken to mean that a healthy diet is unimportant. There are many dietary factors, including vitamins, antioxidants, salt, and sugar, among others, that may play a role in CHD and other diseases.

Ten years ago it was almost malpractice not to endorse estrogen
[for hormone replacement therapy]. Now the bubble has burst.

—*Dr. Isaac Schiff, Massachusetts General Hospital, quoted in
the July 22, 2002 issue of* Time *magazine*

In the words of a *New York Times* July 14, 2002 editorial,
"These have not been good times for established medical practice." It was a great shock to many postmenopausal American
women to learn that their hormone replacement therapy not
only wasn't reducing their risk of cardiovascular disease as
previously thought, but actually was elevating it. Other recent
reversals in conventional medical wisdom would include a
finding that arthroscopic surgery was, in fact, no more effective
than placebo surgery (only small incisions were made), and
that by avoiding dietary fat—particularly saturated fat—
Americans may have been producing (not avoiding) an epidemic of obesity.[1]

Reversals in conventional wisdom seem to happen particularly frequently in the field of medicine and health. In Woody
Allen's movie "Sleeper," upon awakening 200 years in the future the Allen character learns that science has proven that
deep fried foods and cigarettes are actually healthy for you.
That particular reversal hasn't happened yet, but it no longer
seems quite as preposterous, given the ones that have actually
occurred. Why do reversals of conventional wisdom occur so
frequently in the medical field?

There are probably many reasons for such reversals that ultimately relate to the great commercial stakes involved in many
medical studies and the need of drug companies to find new

306 markets and cure new diseases—which may or may not exist. Undaunted by the recent reversal in the conventional wisdom on hormone replacement therapy for women, one large drug conglomerate has been promoting its new product to treat "andropause," the male equivalent of menopause, for older males having reduced testosterone levels. But there is little evidence that "andropause" actually exists, aside from the natural gradual reduction in some physical abilities and appetites that occur as we age. Nor does the lack of scientific studies on the relationship between testosterone levels and the underlying symptoms seem to be a problem for those involved in this marketing effort.[2]

One specific reason why reversals occur in medical studies is that despite the recognized need to conduct medical studies in a double-blind fashion, in many cases the studies become unblinded, and hence biased, as explained in chapter 8. Furthermore, owing to the commercial interests of a study's sponsor, or the belief that studies showing positive results are more "worthy," sometimes studies showing negative results simply remain unpublished. Finally, there is the issue of what constitutes a positive result. Research in the medical field (along with that in social sciences and education), often uses a weak criteria for statistical significance of $p < 0.05$, according to which a chance result should occur only once in 20 times. Some medical researchers have begun to realize that this statistical criterion is insufficient, and have recommended the more stringent $p < 0.001$.[3] But even that test is less stringent than what is required in most experiments in physical science.

The greatest failing in the way statistics is used in many studies lies not in the choice of what constitutes a statistically significant result, however, but in a lack of appreciation for the bias involved when we make an "informed choice" when analyzing data from an experiment. The pitfalls of making informed choices—first seeing a pattern in the data, and then defining a statistical test that asks what the chances are for that particular pattern to appear—were discussed in several places in this book, especially chapter 2.

307   This book has taken a critical look at eight specific ideas and attempted to rate their credibility, based on the weight of the evidence on each side. A summary of my ratings is given at the end of this chapter, where zero snowflakes means that there is a reasonable degree of confidence the idea is true and four snowflakes means that there is no credible evidence. Please remember that these ratings are subjective and reflect not on the "craziness" of each idea, but rather on how well supported each one seems to be based on the evidence presented on each side. Being a scientist who tries to keep an open mind, I might change my mind drastically on some of them if new evidence should come to light.

Lastly, readers may be interested in what of possible "new evidence" has appeared since the publication of *Nine Crazy Ideas in Science*, the prequel to the present book. Have I changed my mind regarding any of the "cuckoo" ratings I assigned there? In some cases new evidence actually existed at the time I made my judgments, but I was not fully aware of it. One such case concerned the controversial idea that AIDS is not caused by HIV, which I rated as "3 cuckoos," meaning highly unlikely. Although I wouldn't change that rating now, I should have also included a discussion of the less extreme view that the development of AIDS requires infection with HIV in addition to other cofactors. The existence of such cofactors would allow some individuals who are infected with the virus to remain AIDS-free.[4] This theory, I believe, is much more credible than the one I discussed.

With respect to the idea of faster-than-light particles, known as tachyons, and the suggestion that the electron neutrino is a tachyon, the experiments remain to be carried out. A more precise measurement of the mass of the electron neutrino may be made in the coming years, which might serve as one such test. Other experiments involving cosmic rays also remain to be done. (Tachyons have the weird property of having an imaginary rest mass.) For completeness, the earlier chapter on faster-than-light (FTL) travel should have also discussed other ways that FTL speeds could be achieved. One such way involves

308 making use of space–time distortions in general relativity to allow FTL speeds—or the "warp drive" of science fiction. Although this method remains a theoretical possibility, it would require the existence of "exotic" kinds of matter that have not been observed.[5] I probably should have also discussed experiments conducted at Princeton in which wave packets were created whose "group velocity" (the speed of the maximum of the pulse) traveled faster than light.[6] Most scientists, however, do not believe that this demonstration involved the transmission of information or energy at FTL speed.

New developments have been reported relevant to several other topics treated previously. The idea that the Big Bang is just one of an endless number of cycles and that our universe is infinitely old has received new impetus from a theory by physicists Turok and Steinhardt.[7] Unlike the conventional Big Bang theory, in this new version there are no singularities, i.e., the density and temperature of the universe remain finite at the Big Crunch marking the end of each cycle. In this respect, Turok and Steinhardt's theory resembles the quasi-steady-state theory of an oscillating universe. Although this new theory involves eleven dimensions, of which we can sense only four, it may eventually be testable once detectors are built that can observe gravitational waves.

Possible new developments have also been reported concerning time travel. Ronald Mallet, a University of Connecticut physicist who has done theoretical work on black holes, is attempting to build a time machine and his efforts have gotten some attention in the popular press.[8] I wish Mallet luck in his endeavors, although I wouldn't invest in it if he were seeking funding. My opinion of the feasibility of time travel remains at "2 cuckoos"—meaning possible, though *very* unlikely. Lastly, the controversial idea that sun exposure is, on balance, good for you has received additional support recently by some research that indicates ultraviolet-B radiation reduces the risk of about a dozen types of cancer.[9]

In table 10.1 I summarize my ratings for the eight ideas discussed in this book.

309 **Table 10.1**

Flakiness Ratings for the Eight Ideas Discussed in This Book

Idea	Rating
Homosexuality is primarily innate.	0 flakes
Intelligent design is a viable scientific alternative to evolution.	3 flakes
People are getting smarter (dumber).	1(2) flakes
The mind can influence external matter by thought alone.	4 flakes
We shouldn't worry (too much) about global warming now.	1 flake
Complex life is very rare in the universe.	2 flakes
A sugar pill can cure (or sicken) you.	0 flakes
High cholesterol is not worth worrying about.	2 flakes

Notes

Chapter One. Introduction

    1. J. L. Freedman, *Media Violence and Its Effect on Aggression: Assessing the Scientific Evidence* (Toronto: University of Toronto Press, 2002).

    2. The degree of correlation between two quantities is measured by their correlation coefficient, which is a number that ranges between plus one (perfectly correlated) and minus one (perfectly anticorrelated). If two positive quantities are perfectly correlated, the size of one will be found to increase whenever the size of the other increases. If they are perfectly anticorrelated, when one goes up the other goes down. Thus, in a given location the sales tax on an item is perfectly correlated with its price. With perfectly anticorrelated quantities the correlation works in the reverse direction: whenever one is large the other is always small. A pair of quantities is said to be uncorrelated (zero correlation coefficient) if the size of one is completely independent of the size of the other.

    3. According to Freedman, far fewer than half the studies have found a causal connection between media violence and aggression or crime. Here are two possible ways A and B could be correlated without A causing B. First, maybe kids who are more aggressive like to watch more violent TV programming, so the cause and effect work in the reverse direction. Second, maybe parents who don't supervise what their children watch permit them to watch more violent programming, and it's that lack of parental supervision, not the content of the shows, that causes later problems with violence.

    4. It may not be entirely irrelevant that the United States Motion Pictures Association funds Freedman's research.

Chapter Two. Is Homosexuality Primarily Innate?

    1. For the lower estimate, see the National Health and Social Life Survey, 1992: www.rwjf.org/publications/publicationsPdfs/library/oldhealth/chap11.htm. For the higher estimate, see www.kinseyinstitute.org/resources/bib-homoprev.htm.

312   2. A. N. Groth and H. J. Birnbaum, Adult sexual orientation and attraction to underage persons, *Archives of Sexual Behavior* 7(3): 175–81 (1978). Also, see information provided by the American Psychological Association at www.apa.org/pubinfo/orient.html.

3. R. Green. *The Sissy Boy Syndrome, and the Development of Homosexuality* (New Haven, CT: Yale University Press, 1987).

4. F. L. Whitam and M. Zent, A cross-cultural assessment of early cross gender behavior and familial factors in male homosexuality, *Archives of Sexual Behavior* 13:427–41 (1984).

5. A. Kinsey, W. Pomeroy, and C. Martin, *Sexual Behavior in the Human Male* (Philadelphia: W. B. Saunders, 1948); A. Kinsey, W. Pomeroy, C. Martin, and P. Gebhard, *Sexual Behavior in the Human Female* (Philadelphia: W. B. Saunders, 1953).

6. H. E. Adams, L. W. Wright Jr., and B. A. Lohr, Is homophobia associated with homosexual arousal? *Journal of Abnormal Psychology* 105(3): 440–45 (1996).

7. American Psychiatric Association, *Diagnostic and Statistical Manual of Mental Disorders*, 4th ed. (DSM-IV) (Washington, DC: American Psychiatric Association, 1994).

8. E. Balaban, M.-A. Teillet, and N. LeDouarin, Application of the quail-chick chimeric system to the study of brain development and behavior, *Science* 241:1339–42 (1990).

9. C. Burr, *A Separate Creation: The Search for the Biological Origins of Sexual Orientation* (New York: Hyperion, 1996).

10. J. C. Gonsiorek, R. L. Sell, and J. D. Weinrich, Definition and measurement of sexual orientation, *Suicide and Life-Threatening Behavior* 25 (suppl.): 40–51 (1995). For a summary of the literature, see www.kinseyinstitute.org/resources/bib-homoprev.htm.

11. M. Diamond, Homosexuality and bisexuality in different populations, *Archives of Sexual Behavior* 22(4): 291–310 (1993).

12. B. Bagemihl, *Biological Exuberance: Animal Homosexuality and Natural Diversity*, illustrated by J. Megahan (New York: St. Martin's Press, 1999).

13. See J. M. Bailey, D. Bobrow, M. Wolfe, and S. Mikach, Sexual orientation of adult sons of gay fathers, *Developmental Psychology* 31:124–29 (1995); F. W. Bozett, Children of gay fathers, in *Gay and Lesbian Parents*, ed. F. W. Bozett (New York: Praeger, 1987), 39–57; J. S. Gottman, Children of gay and lesbian parents, in *Homosexuality and Family Relations*, ed. F. W. Bozett and M. B. Sussman (New York: Harrington Park Press, 1991), 177–96.

14. H. A. Halem et al., Medial preoptic/anterior hypothalamic lesions induce a female typical profile of sexual partner preference in male ferrets, *Hormones and Behavior* 30:514–27 (1996).

313   15. K. Zucker et al., Psychosexual development of women with congenital adrenal hyperplasia, *Hormones and Behavior* 30:300–18 (1996).

16. L. Ellis, H. Hoffman, and D. M. Burke, Sex, sexual orientation and criminal and violent behavior, *Personality and Individual Differences* 11:1207–12 (1990).

17. Simon LeVay, in his book *Queer Science* (Cambridge, MA: MIT Press, 1996), notes that research shows that after adjusting for differences in average body weight, men's brains are 7–8 percent heavier than women's brains (p. 139).

18. K. M. Bishop and D. Wahlsten, Sex differences in the human corpus callosum: myth or reality? *Neuroscience and Behavioral Reviews* 21:581–601 (1997); M. S. Lasco, T. J. Jordan, M. A. Edgar, C. K. Petito, and W. Byne, A lack of dimorphism of sex or sexual orientation in the human anterior commissure, *Brain Research* 936(1–2): 95–98 (2002).

19. R. Gorski, R. Harlan, C. Jacobsen, J. Shryne, and A. Southam, Evidence for a morphological sex difference within the medial preoptic area of the rat brain, *Journal of Comparative Neurology* 193:529–39 (1980).

20. S. LeVay, A difference in hypothalamic structure between heterosexual and homosexual men, *Science* 253:1034–37 (1991).

21. W. Byne, S. Tobet, L. A. Mattiace, M. S. Lasco, E. Kemether, M. A. Edgar, S. Morgello, M. S. Buchsbaum, and L. B. Jones, The interstitial nuclei of the human anterior hypothalamus: and investigation of variation with sex, sexual orientation and HIV status, *Hormones and Behavior* 40(2): 86–92 (2001).

22. J. M. Bailey and R. C. Pillard, A genetic study of male sexual orientation, *Archives of General Psychiatry* 48:1089–96 (1991).

23. J. M. Bailey and N. G. Martin, A twin registry study of sexual orientation (poster presentation at the 21st annual meeting of the International Academy of Sex Research, Provincetown, MA, 1995).

24. D. H. Hamer and P. Copeland, *The Science of Desire: The Search for the Gay Gene and the Biology of Behavior* (New York: Simon & Schuster, 1994).

25. D. Hamer, S. Hu, V. Magnuson, N. Hu, and A. Pattatucci, A linkage between DNA markers on the X chromosome and male sexual orientation, *Science* 261:321–27 (1993).

26. A.M.L. Pattatucci, and D. H. Hamer, Development and familiality of sexual orientation in females, *Behavior Genetics* 25(5): 407–20 (1995).

27. The follow-up study by Hamer's group was S. Hu, Linkage between sexual orientation and chromosome Xq28 in males but not in females, *Nature* 11:248 (1995); see also G. Rice, C. Anderson, N. Risch, and G. Ebers, Male homosexuality: absence of linkage to microsatellite markers at Xq28, *Science* 284:665 (1999); A. R. Sanders et al., Poster presentation

314 149 (annual meeting of the American Psychiatric Association, Toronto, Ontario, 1998).

28. J. M. Bailey et al., A family history study of male sexual orientation using three independent samples, *Behavior Genetics* 29:79 (1999).

29. T. Kitamoto, Conditional disruption of synaptic transmission induces male-male courtship in *Drosophila*, *Proceedings of the National Academy of Sciences of the United States of America* 99:13232–37 (2002).

30. See www.gallup.com/Poll/releases/pr010604.asp.

31. Ibid.

32. G. Easterbrook, Inconceivable, *The New Republic*, 23 November 1998.

Chapter Three. Is Intelligent Design a Scientific Alternative to Evolution?

1. G. Bishop, The religious worldview and American beliefs about human origins, *The Public Perspective*, August 1998.

2. The survey included Israel, which is not a Christian country. The other countries in the survey besides the United States were Northern Ireland, Philippines, Ireland, Poland, Italy, New Zealand, Austria, Norway, Great Britain, Netherlands, West Germany, Russia, Slovenia, Hungary, and East Germany.

3. U. Sayin and A. Kence, Islamic scientific creationism: a new challenge in Turkey, *National Center for Science Education Reports* 19:18–29 (1999).

4. R. Dawkins, *The Selfish Gene* (New York: Oxford University Press, 1976).

5. The judge found Scopes guilty, and fined him $100, but the case was overturned by the Tennessee Supreme Court on a technicality: only a jury could then impose fines above $50. This ruling prevented an appeal to the Supreme Court, although the Court has ruled on several similar cases since then, as discussed later.

6. K. Miller, *Finding Darwin's God: A Scientist's Search for Common Ground between God and Evolution* (New York: Harper Collins, 1999).

7. Kenneth Miller has expressed the view that a God who would create the world less than 10,000 years ago as the Bible says, and merely give it the appearance of great age would have to be a charlatan. But it is impossible to know what God's purposes might be, if indeed we were to make such an assumption. God may have wanted to test our faith, or perhaps create a mature universe. (I don't believe either of these ideas is true, but it is dangerous to impute motives to God.)

8. S. I. Dutch, Religion as belief versus religion as fact, *Journal of Geoscience Education* 50(2): 137–44 (March 2002).

315   9. M. Behe, *Darwin's Black Box: The Biochemical Challenge to Evolution* (New York: Simon & Schuster, 1996).

10. R. Dawkins, *Climbing Mount Improbable* (New York: W.W. Norton, 1996).

11. S. J. Gould and N. Eldredge, Punctuated equilibrium comes of age, *Nature* 366:223–27 (1993).

12. In the punctuated equilibrium theory, however, it is assumed that biological species split off not in a single generation but merely in a time that is short on a geological timescale. In reality, the distinction between conventional Darwinian evolution, with its emphasis on gradual change over time, and the theory of punctuated equilibrium is merely a question of the timescale over which change occurs.

13. Genetic materials can be swapped, or switched on or off, whole viral genes can be inserted, introns/extrons can swap places, the number of structures can be controlled by some other gene, centromeres can change position, arms can become longer or shorter, winding can become tight or loose, chromosomes can fuse, shuffle, or undermethylate, etc.

14. A. W. Crompton and F. A. Jenkins, Origin of mammals, in *Mesozoic Mammals: The First Two Thirds of Mammalian History*, ed. Z. Kielan-Jaworowska, J. G. Eaton, and T. M. Brown (Berkeley: University of California Press, 1979).

15. S. B. Carroll, S. D. Weatherbee, and J. A. Langeland, Homeotic genes and the regulation and evolution of insect wing number, *Nature* 375:58–61 (1995).

16. C. Zimmer, Crystal balls, *Natural History,* April 2002.

17. R. F. Doolittle, The evolution of vertebrate blood coagulation: a case of yin and yang, *Thrombosis and Haemostasis* 70:24–28 (1993).

18. B. Carter, Large number coincidences and the anthropic principle, in *Confrontation of Cosmological Theories with Observational Data,* ed. M. S. Longair (Boston: D. Reidel, 1974), 291–98.

19. J. D. Barrow and F. J. Tipler, *The Anthropic Cosmological Principle* (Oxford, U.K.: Clarendon Press, 1986).

20. S. Weinberg, Life in the universe, *Scientific American,* October 1994, 49.

Chapter Four. Are People Getting Smarter or Dumber?

1. Psychologist Robert Butterworth, quoted in the *Washington Post,* 23 August 2001, calls it the "lottery fantasy syndrome."

2. W. Northcutt, *The Darwin Awards: Evolution in Action* (New York: Penguin Putnam, 2002).

316   3. H. Gardner, *Multiple Intelligences: The Theory in Practice* (New York: Basic Books, 1993).

4. R. Rothstein, *The Way We Were: The Myths and Realities of America's Student Achievement* (Washington: The Brookings Institute, Century Foundation Press, 1998).

5. D. Kearns, An education recovery plan for America, *Phi Delta Kappan*, April 1988, 565–70.

6. V. Packard, Are we becoming a nation of illiterates? *Reader's Digest*, April 1974, 81–85.

7. C. C. Walcutt, *Tomorrow's Illiterates: The State of Reading Instruction Today* (Boston: Atlantic Monthly Press, 1961), xiii–xvi.

8. Interview with Arthur Bestor: What went wrong with U.S. schools, *U.S. News and World Report*, 24 January 1958, 68–77.

9. B. Fine, *Our Children Are Cheated* (New York: Holt, 1947).

10. B. Fine, Ignorance of U.S. history shown by college freshmen, *New York Times*, 4 April 1943.

11. D. C. Berliner, Educational reform in an era of disinformation (paper presented at the annual meeting of the American Association of Colleges of Teacher Education, San Antonio, TX, February 1992), 54.

12. Rothstein, *The Way We Were*.

13. D. P. Hayes, L. T. Wolfer, and M. F. Wolfe, Schoolbook simplification and its relation to the decline in SAT-verbal scores, *American Educational Research Journal* 33:1–18 (1996).

14. J. Hubisz, *The Physics Teacher*, May 2001; also, see www.psrc-online. org/curriculum/hubisz.htm.

15. Hubisz, private communication.

16. T. O'Banion, *A Learning College for the 21st Century* (Phoenix, AZ: Oryx Press, 1997).

17. Most people are aware that the dinosaurs died out 65 million years ago, so that humans and dinosaurs missed each other by around 60 million years. But local residents of the Paluxy river valley in Texas created a hoax in the 1930s by carving some human footprints in among dinosaur tracks found in fossilized limestone. Many people wishing to validate their creationist beliefs still accept this fraudulent evidence for human–dinosaur coexistence.

18. *Harvard University Gazette*, 29 May 1997.

19. *Minds of Our Own: Can We Believe Our Eyes?* (South Burlington, VT: Annenberg Corporation for Public Broadcasting Multimedia, 1987).

20. R. W. Howard, Preliminary real-world evidence that average human intelligence really is rising, *Intelligence* 27(3): 235–50 (1999).

317   21. For the data on physics degrees, see www.aip.org/statistics/index.htm. Also, see National Science Foundation, *Science and Engineering Indicators* (Washington, DC: Government Printing Office, 2000). Downloadable from www.nsf.gov/sbe/srs/seind00/start.htm.

22. The reason why many people believe that the seasons are caused by variations in the Earth's distance to the sun during its orbit is that science teachers emphasize that the path of the Earth around the sun is an ellipse. But teachers sometimes neglect to mention that this ellipse is so close to being a circle that you couldn't tell it isn't a circle by looking at it.

23. W. Williams, Are we raising smarter children today? School and home-related influences on IQ, in *The Rising Curve: Long-Term Changes in IQ and Related Measures*, ed. U. Neisser (Washington, DC: American Psychological Book Association, 1998).

24. W. Wirtz and H. Howe II, *On Further Examination: Report of the Advisory Panel on the Scholastic Aptitude Test Score Decline* (Princeton, NJ: Educational Testing Service, 1977).

25. The College Board, www.collegeboard.com.

26. National Center for Education Statistics, *NAEP 1999 Trends in Academic Progress: Three Decades of Student Performance*. Downloadable from nces.ed.gov/nationsreportcard/about/trend.asp.

27. National Science Foundation, *Science and Engineering Indicators*.

28. For the 1990 and 2001 polling data, see www.gallup.com/poll/releases/pr010608.asp. For the 1978 polling data, see *The Gallup Poll* (1978).

29. A. Greeley, From here to the hereafter, *San Jose Mercury News*, 17 January 1987, C-1.

30. To learn how this lunacy can be debunked easily, see P. Plait, Men on the moon as a matter of fact, *Space Illustrated*, Fall 2001, 24–25.

31. J. C. Burnham, *How Superstition Won and Science Lost* (New Brunswick, NJ: Rutgers University Press, 1987).

32. E. E. Levitt, Superstitions: twenty-five years ago and today, *American Journal of Psychology* 65(3): 443–49 (1952).

33. D. W. Forrest, *Francis Galton: The Life and Work of a Victorian Genius* (New York: Taplinger, 1974). Also, see www.tld.jcu.edu.au/hist/stats/reading.html.

34. For a meta-analysis of the relevant studies, see H. M. Chipuer, M. Rovine, and R. Plomin, LISREL modeling: genetic and environmental influences on IQ revisited, *Intelligence* 14:11–29 (1990).

35. For a summary of the results from various studies see R. Lynn, The decline of genotypic intelligence, which appears in Neisser (ed.), *The Rising Curve* (see note 23).

318   36. I. W. Waldman, Problems in inferring dysgenic trends for intelligence, in Neisser (ed.), *The Rising Curve* (see note 23).

37. Lynn, The decline of genotypic intelligence.

38. J. R. Flynn, IQ gains over time: toward finding the causes, in Neisser (ed.), *The Rising Curve* (see note 23).

39. Ibid.

40. Ibid.

41. P. A. Carpenter, M. A. Just, and P. Shell, What one intelligence test measures, *Psychological Review* 97:409 (1990).

42. Flynn, IQ gains over time.

43. Ibid.; and Flynn, private communication.

44. S. Cahan and N. Cohen, Age versus schooling effects on intelligence development, *Child Development* 60:1239–49 (1989).

45. E. A. Cocodia et al., Evidence that rising intelligence is impacting in formal education, in press.

46. Howard, Evidence that average human intelligence really is rising.

47. J. Rosenau and W. M. Fagen, A new dynamism in world politics: increasingly skillful individuals? *International Studies Quarterly* 41:655–86 (1997).

48. Educational Testing Service, Princeton, NJ, www.ets.org.

49. W. Dickens and J. R. Flynn, Heritability estimates vs. large environmental effects: the IQ paradox resolved, *Psychological Review* 108(2): 346–69 (2001).

50. S. J. Gould, *Full House: The Spread of Excellence from Plato to Darwin* (New York: Harmony Books, 1996).

Chapter Five. Can We Influence Matter by Thought Alone?

1. R. Targ and J. Katra, *Miracles of the Mind: Exploring Nonlocal Consciousness and Spiritual Healing* (Novato, CA: New World Library, 1999).

2. R. Ehrlich, *Why Toast Lands Jelly-Side Down* (Princeton, NJ: Princeton University Press, 1997).

3. See www.brainfingers.com/index.html.

4. See footnotes 7–18 in R. G. Jahn et al., Correlations of random binary sequences with pre-stated operator intention: a review of a 12-year program, *Journal of Scientific Exploration* 11(3): 345–67 (1997).

5. N. Bohr, *Atomic Theory and the Description of Nature* (Cambridge, U.K.: Cambridge University Press, 1961).

6. Targ and Katra, *Miracles of the Mind.*

319   7. D. Radin, *The Conscious Universe: The Scientific Truth of Psychic Phenomena* (New York: Harper Collins, 1997); and see also Targ and Katra, *Miracles of the Mind*.

8. P. Anderson, *Physics Today,* October 1991, 146.

9. Ibid.

10. Although the late astronomer Carl Sagan debunked beliefs about the paranormal throughout his career, in his 1995 book *The Demon Haunted World* (New York: Ballantine Books, 1997), he suggested three claims in the ESP field that deserve serious study. In Sagan's words, these three claims were that (1) by thought alone humans can (barely) affect random number generators, (2) people under mild sensory deprivation can receive thoughts or images projected at them, and (3) young children sometimes report details of a previous life, which upon checking turn out to be accurate, and which they could not have known about in any way other than reincarnation.

11. J. B. Rhine et al., *Parapsychology from Duke to FRNM* (Durham, NC: The Parapsychology Press, 1965). For a good overview of Rhine's studies on psychokinesis, see www.williamjames.com/Science/PK.htm.

12. See Randi's web site: www.randi.org/research.

13. W. Roll, Physical and psychological aspects of recurrent psychokinesis (Dinsdale Prize Lecture at the 21st annual meeting of the Society for Scientific Exploration, Charlottesville, VA, 29–31 May 2002).

14. K. Wiesenfeld and F. Moss, *Nature* 373:33–36 (1995); L. Gammaitoni, P. Hanggi, and P. Jung, *Reviews in Modern Physics* 70:223–88 (1998).

15. For information about the journal see http://www.scientificexploration.org/jse.html.

16. R. G. Jahn et al., Correlations of random binary sequences with pre-stated operator intention: a review of a 12-year program, *Journal of Scientific Exploration* 11(3): 345–67 (1997).

17. According to Dr. Jahn, there are a number of pragmatic and psychological reasons for not continuously monitoring operators. Moreover, extensive tests have been conducted of the REG devices to show that no distortion of the results can be created by subjecting them to touching, thumping, shaking, heating, or electromagnetic interference.

18. Jahn, informal communication.

19. Jahn et al., Correlations of random binary sequences.

20. B. J. Dunne and R. G. Jahn, Consciousness and anomalous physical phenomena, Princeton Engineering Anomalies Research, Technical Note PEAR 95004, May 1995.

320   21. Ibid.

22. Ibid.

23. Ibid.

24. Ibid.

25. Jahn et al., Correlations of random binary sequences.

26. Y. Dobyns, High bitrate REG experiments (paper presented at the 21st annual meeting of the Society for Scientific Exploration, Charlottesville, VA, 29–31 May 2002).

27. B. J. Dunne and R. G. Jahn, Experiments in remote human/machine interaction, *Journal of Scientific Exploration* 6(4): 311–32 (1992).

28. R. D. Nelson, Wishing for good weather: a natural experiment in group consciousness, *Journal of Scientific Exploration* 11:47–58 (1997).

29. D. Radin and R. D. Nelson, Meta-analysis of mind-matter interaction experiments: 1959 to 2000. See www.boundaryinstitute.org/articles/ rngma.pdf. This paper will appear in W. Jonas and C. Crawford (eds.), *Science and Spiritual Healing: A Critical Review of Research on Spiritual Healing, Energy Medicine and Intentionality* (London: Harcourt Health Sciences, in press).

30. Ibid.

31. Ibid.

32. Ibid.

Chapter Six. Should You Worry about Global Warming?

1. J. T. Houghton, Y. Ding, D. J. Griggs, M. Noguer, P. J. van der Linden, and D. Xiaosu (eds.), *Climate Change, 2001: The Scientific Basis.* The Third Assessment Report of the Intergovernmental Panel on Climate Change (IPCC) (Cambridge, U.K.: Cambridge University Press, 2001).

2. R. Lindzen, Absence of scientific basis, *Research & Exoloration* 9(2): 191–200 (1993).

3. D. Grossman, Dissent in the mainstream, *Scientific American,* November 2001, 1–3.

4. There are many possible terms that could be applied to those on the other side of the debate from the skeptics, but none seems quite appropriate. For example, the opposite of a skeptic is a believer, but that term seems more appropriate to a religious or political issue than a scientific debate. We could call people who believe steps should be taken now worriers, but that term seems too similar to the pejorative alarmists. The term, concerned, would probably be welcome by those on that side of the debate, but that

321 term implies (perhaps unfairly) that the other side of the debate is unconcerned about the environment.

5. See www.gallup.com/poll/Releases/Pr010409.asp.

6. S. G. Philander, *Is the Temperature Rising? The Uncertain Science of Global
Warming* (Princeton, NJ: Princeton University Press, 1998).

7. The IPCC assumes average CO_2 residence times in the atmosphere of
about a century, but skeptics such as Peter Dietz believe that the true time is
only about a tenth that long; see www.john-daly.com/forcing/moderr.htm.

8. Shukla, private communication.

9. P. J. Michaels and R. C. Balling Jr., *Satanic Gases: Clearing the Air about
Global Warming* (Washington, DC: CATO Institute, 2000).

10. Summers, private communication.

11. Houghton et al., *Climate Change, 2001*.

12. B. Santer et al., Accounting for the effects of volcanoes and ENSO in
comparisons of modeled and observed temperature trends, *Journal of
Geophysical Research—Atmospheres* 106:28033–59 (2001).

13. Houghton et al., *Climate Change, 2001*.

14. H. Fischer et al., Ice core records of atmospheric CO_2 around the last
three glacial terminations, *Science* 283(5408): 1712 (1999).

15. *Science,* 7 December 2001, articles on pp. 2049, 2109, 2130, and 2149.

16. E. Friis-Christensen and K. Lassen, Length of the solar cycle: an indicator of solar activity closely associated with climate, *Science* 254:698–700
(1991).

17. S. F. Singer, *Hot Talk, Cold Science: Global Warming's Unfinished Debate*
(Oakland, CA: The Independent Institute, 1999).

18. To judge the validity of this smoothing procedure, imagine someone
claiming to find a close correlation between crime rates and the annual unemployment figures, but only when the unemployment rate is averaged
with the figures two years behind and two years *ahead of* the present year.

19. P. A. Thejll and K. Lassen, Solar forcing of the Northern Hemisphere
land air temperature: new data, *Journal of Solar-Terrestrial Physics* 13:1207–13
(2000).

20. U. Neff et al., Strong coherence between solar variability and the
monsoon in Oman between 9 and 6 kyr ago, *Nature* 411:290–93 (2001).

21. Michaels and Balling, *Satanic Gases.*

22. P. J. Michaels et al., Observed warming in cold anticyclones, *Climate
Research* 14:1–6 (2000).

23. Houghton et al., *Climate Change, 2001.*

24. Michaels and Balling, *Satanic Gases.*

322   25. Houghton et al., *Climate Change, 2001.*

26. Singer, *Hot Talk, Cold Science.*

27. K. Deffeyes, *Hubbert's Peak: The Impending World Oil Shortage* (Princeton, NJ: Princeton University Press, 2001).

28. James Hansen, www.giss.nasa.gov/edu/gwdebate.

29. S. Schneider, What is "dangerous" climate change? *Nature* 411:17–19 (2001).

30. Interview with Jonathan Schell, *Discover*, October 1989, 45–48.

31. S. Schneider, *Global Warming: Are We Entering the Greenhouse Century?* (San Francisco: Sierra Club Books, 1989).

32. Schneider, private communication.

33. T.M.L. Wigley and S.C.B. Raper, Interpretation of high projections for global-mean warming, *Science* 293:451–54 (2001).

34. Vital Statistics of the United States, U.S. Department of Commerce. Various years. *Statistical Abstract of the United States* (Washington, DC: Government Printing Office).

35. W. R. Keatinge and G. C. Donaldson, Mortality related to cold and air pollution in London after allowance for effects of associated weather patterns, *Environmental Research* 86:209–16 (2001).

36. Houghton et al., *Climate Change, 2001.*

37. Ibid.

38. Singer, *Hot Talk, Cold Science.*

39. B. C. Douglas and W. R. Peltier, The puzzle of global sea-level rise, *Physics Today*, March 2002, 35–40.

40. T. Karl, R. W. Knight, and N. Plummer, Trends in high-frequency climate variability in the twentieth century, *Nature* 377:217–20 (1995).

41. R. Mendelsohn and J. E. Neuman (eds.), *The Impact of Climate Change on the United States Economy* (Cambridge, U.K.: Cambridge University Press, 1999).

42. R. B. Myneni et al., Increased plant growth in the northern high latitudes, from 1981 to 1991, *Nature* 386:698–702 (1997).

43. S. H. Wittwer, *Food, Climate, and Carbon Dioxide* (Boca Raton, FL: CRC Press, 1995).

44. S. P. Long, Modification of the response of photosynthetic reproductivity to rising temperature by atmospheric CO_2 concentrations: has its importance been underestimated? *Plant Cell and Environment* 14:729–39 (1991).

45. Mendelsohn and Neuman (eds.), *Impact of Climate Change.*

46. Ibid.

47. Michaels and Balling, *Satanic Gases.*

48. Houghton et al., *Climate Change, 2001.*

323   49. Ibid.

50. Michaels and Balling, *Satanic Gases*.

51. E. V. Mielczarek, *Iron, Nature's Universal Element: Why People Need Iron and Animals Make Magnets* (New Brunswick, NJ: Rutgers University Press, 2000).

52. P. W. Boyd et al., A mesoscale phytoplankton bloom in the polar Southern Ocean stimulated by iron fertilization, *Nature* 407:695–702 (2000).

53. S. Chisholm, Oceanography: stirring times in the southern ocean, *Nature* 407:685–86 (2000).

54. J. Houghton, *Global Warming: The Complete Briefing*, 2d ed. (New York: Cambridge University Press, 1997).

Chapter Seven. Is Complex Life in the Universe Very Rare?

1. See www.gallup.com/poll/fromtheed/ed9710.asp.

2. C. Sagan, *Cosmos* (New York: Random House, 1980).

3. T. Kuiper and G. D. Brin, *Extraterrestrial Civilization* (College Park, MD: American Association of Physics Teachers, 1989).

4. Michael Shermer notes that based on historical trends, the average Earthly civilization has lasted around 400 years. On that basis he suggests that there might be only two or three intelligent civilizations in the galaxy at any given time. See M. Shermer, Why ET hasn't called, *Scientific American*, August 2002, 33.

5. P. D. Ward and D. Brownlee, *Rare Earth: Why Complex Life Is Uncommon in the Universe* (New York: Springer-Verlag, 1999).

6. S. A. Kauffman, Self-replication: even peptides do it, *Nature* 382:496–97 (1996).

7. T. Gold, *The Deep Hot Biosphere* (New York: Springer-Verlag, 1999).

8. S. Arrhenius, *Worlds in the Making* (New York: Harper & Roe, 1908).

9. F. Hoyle and N. C. Wickramasinghe, *The Theory of Cosmic Grains* (Dordrecht: Kluwer, 1990); C. Wickramasinghe, The long road to panspermia, *Astronomy Now* 16:57–60 (April 2002); also, see www.panspermia.org.

10. D. Schwartzman, M. McMenamin, and T. Volk, Did surface temperatures constrain microbial evolution? *BioScience* 43:390–93 (1993).

11. Ward and Brownlee, *Rare Earth*.

12. Astronomers count all elements with atomic number greater than 2 (helium) as being "heavy."

13. An up-to-date catalog of confirmed extrasolar planets can be found at www.obspm.fr/encycl/catalog.html.

324   14. A. G. Cameron and R. M. Canup, The giant impact occurred during the Earth accretion, *Lunar and Planetary Science Conference* 29:1062 (1998).

15. Ward and Brownlee, *Rare Earth*.

16. J. L. Kirschvink, A paleogeographic model for Vendian and Cambrian time, in *The Proterozoic Biosphere: A Multidisciplinary Study*, ed. J. W. Schopf, C. Klein, and D. Des Maris (Cambridge, U.K.: Cambridge University Press, 1992), 567–81.

17. W. Brandner, Binary surveys: constraining the number of planetary systems (paper presented at the Workshop on Planetary Formation in Binary Environment, Stony Brook, NY, 16–18 June 1996).

18. This estimate is based on a calculation by cosmic ray physicist John Bieber at the University of Delaware. Bieber notes that the maximum shielding at sea level provided by the Earth's magnetic field is around a factor of two. If the atmosphere were thicker by around 10 percent we would also gain the same factor of two extra shielding. (The total shielding provided by the atmosphere reduces the cosmic ray flux to a mere 0.02 percent of its value without an atmosphere.)

19. The exclusion of the Jovian planets is reasonable based on their being gas giants, primarily consisting of hydrogen and helium, from which it is difficult to form a large moon.

20. H. Hartman and C. P. McKay, Oxygenic photosynthesis and the oxidation-state of Mars, *Planetary and Space Science* 43(1–2): 123ff. (1995).

21. Quoted in Ward and Brownlee, *Rare Earth*, 214.

22. For information on the detection, see oposite.stsci.edu/pubinfo/PR/2001/38/pr.html.

23. See www.obspm.fr/encycl/catalog.html.

24. The sun-like star known as 55-Cancri has a Jupiter-like planet almost the same distance from the star as Jupiter is from our sun. It also has several close-in Jupiter-size planets, but it is not known whether it also has any smaller Earth-size planets, according to a story in the *Washington Post*, 14 June 2002, W. Harwood, Solar system akin to Earth's is discovered.

25. Ward and Brownlee, *Rare Earth*.

26. Ibid.

27. D. Darling, *Life Everywhere: The Maverick Science of Astrobiology* (New York: Basic Books, 2001).

28. H. Ross, Big bang model refined by fire, in *Mere Creation*, ed. W. A. Demski (Downers Grove, IL: Intervarsity Press, 1998).

29. S. Weinberg, Life in the universe, *Scientific American*, October 1994, 49.

325   30. D. Raup, A kill curve for Phanerozoic marine species, *Paleobiology* 17:37–48 (1991).

31. E. K. Gibson Jr., D. S. McKay, K. Thomas-Keprta, and C. S. Romanek, The case for relic life on Mars, *Scientific American*, December 1997, 58–65; and for more detail, see cass.jsc.nasa.gov/pub/lpi/meteorites/mars_meteorite.html.

32. See web site maintained by Michael Perryman: mperryma@astro.estec.esa.nl.

33. A. K. Dewdney, *Yes, We Have No Neutrons* (New York: John Wiley & Sons, 1997).

34. See chapter 9 in R. Ehrlich, *Nine Crazy Ideas in Science: A Few Might Even Be True* (Princeton, NJ: Princeton University Press, 2001). The reality of tachyons remains uncertain at the time of this writing.

Chapter Eight. Can a Sugar Pill Cure You?

1. A. K. Shapiro and E. Shapiro, The placebo: is it much ado about nothing? in *The Placebo Effect: An Interdisciplinary Exploration*, ed. A. Harrington (Cambridge MA: Harvard University Press, 1997).

2. Editorial—The bottle of medicine, *British Journal of Medicine* 1:149 (1952).

3. W.H.R. Rivers, *The Influence of Alcohol and Other Drugs on Fatigue* (London: Arnold Publishing Co., 1908).

4. Shapiro and Shapiro, The placebo.

5. M. B. Leon, D. S. Baim, J. W. Moses, et al., A randomized blinded clinical trial comparing percutaneous laser myocardial revascularization (using biosense LV mapping) vs. placebo in patients with refractory coronary ischemia (paper presented at a meeting of the American Heart Association, 2000).

6. The use of a double-blind study would be expected in almost any symptom trial and in trials where a physical measurement is important, such as hypertension or obesity, but it is not always expected in trials where the endpoint is wholly objective (e.g., a mortality trial). In such cases, some endpoints may be read by a blinded endpoints committee. Many surgery trials are unblinded, because sham operations bother a lot of people.

7. J. Reich, The effect of personality on placebo response in panic patients, *Journal of Nervous and Mental Diseases* 178:699–702 (1990).

8. D. Joralemon, *Exploring Medical Anthropology* (Boston: Allyn and Bacon, 1998).

326   9. H. K. Beecher, The powerful placebo, *Journal of the American Medical Association* 159:1602–6 (1955).

10. P. Kissel and D. Barrucand, *Placebo and Placebo Effect in Medicine* (Paris: Masson, 1964).

11. S. Wolf, Effects of suggestion and conditioning on the action of chemical agents in human subjects—the pharmacology of placebos, *Journal of Clinical Investigation* 29:100–109 (1950). It should also be noted that ipecac is regarded as a *remedy* for nausea by homeopaths; see www.internethealth library.com/hom-library/ipecac.htm.

12. H. Brody and D. Weismantel, A challenge to core beliefs, written in response to the article by Hrobjartsson and Gotzsche (see note 13), in *Advances in Mind-Body Medicine* 17:296–98 (2001).

13. A. Hrobjartsson and P. Gotzsche, Is the placebo powerless? An analysis of clinical trials comparing placebo with no treatment, *New England Journal of Medicine* 344(21): 1594–99 (2001).

14. F. Galton, Regression towards mediocrity in hereditary stature, *Journal of the Anthropological Institute*, 246–63 (1886).

15. F. E. Anderson, Warts: fact and fiction, *Drugs* 30:368–75 (1985).

16. E. Garfield, *Essays of an Information Scientist: Science Literacy, Policy, Evaluation, and Other Essays* (Philadelphia, PA: Isi Press, 1988), 11.

17. A.H.C. Sinclair-Gieben and D. Chalmers, Evaluation of treatment of warts by hypnosis, *Lancet* 2:480–82 (1959).

18. Shapiro and Shapiro, The placebo.

19. Ibid.

20. Ibid.

21. M.E.P. Seligman, The effectiveness of psychotherapy: the *Consumer Reports* study, *American Psychologist* 50(12): 965–74 (1995).

22. J. Copeland, *Artificial Intelligence: A Philosophical Introduction* (Oxford, U.K.: Blackwell, 1993).

23. M. K. Jacobs, A. Christensen, J. R. Snibbe, et al., A comparison of computer-based versus traditional individual psychotherapy, *Professional Psychology: Research and Practice* 32(1): 92–98 (2001).

24. Copeland, *Artificial Intelligence*.

25. S. Fischer and R. P. Greenberg, *The Limits of Biological Treatments for Psychological Distress: Comparisons with Psychotherapy and Placebo* (Hillsdale, NJ: Lawrence Erlbaum Associates, 1989).

26. See article by Willam Faloon at www.lef.org/magazine/mag2002/apr2002_awsi_01.html.

27. Public Citizen analysis of company reports published in *Fortune* magazine, April 2002.

327   28. See the Industry Intelligence newsletter at www.inpharm.com/intelligence/rbi010699.html.

29. N. Dodman, *Dogs Behaving Badly: An A-to-Z Guide to Understanding and Curing Behavioral Problems in Dogs* (New York: Bantam Books, 1999); also, see www.petplace.com/articles/artShow.asp?artID=3019.

30. Of course, for the placebo effect to work, you would have to believe that the pill you were being given was the real drug or that it might be. For purposes of complete disclosure, a doctor or pharmacist who gave you a box of pills that might be placebos would need to make you aware of that possibility. If the box were taken from a carton containing one box of the real drug and 100 boxes of placebos costing next to nothing, your cost of the drug could be reduced by nearly 99 percent. The placebo effect might be considerably reduced in this case if you knew the low odds of getting the real drug.

31. J. Moncrieff, S. Wessely, and R. Hardy, Meta-analysis of trials comparing antidepressants with active placebos, *British Journal of Psychiatry* 172:227–31 (1998), with commentary by D. Healy, pp. 232–34.

32. J. Rabkin, J. Markowitz, J. Stewart, et al., How blind is blind? Assessment of patient and doctor medication guesses in a placebo-controlled trial of imipramine and phenelzine, *Psychiatry Research* 19:75–86 (1986); J. Margraf, A. Ehlers, W. T. Roth, et al., How "blind" are double-blind studies? *Journal of Consulting and Clinical Psychology* 59:184–87 (1991).

33. D. M. Englehardt, R. A. Margolis, L. Rudorfer, et al., Physician bias and the double-blind, *Archive of General Psychiatry* 20:315–20 (1969).

34. A. Khan, R. M. Leventhal, S. R. Khan, and W. A. Brown, Severity of depression and response to antidepressants and placebo: an analysis of the Food and Drug Administration database, *Journal of Clinical Psychopharmacology* 22:40–45 (2002).

35. I. Kirsch and G. Sapirstein, Listening to Prozac but hearing placebo: a meta-analysis of antidepressant medication, *Prevention & Treatment* 1, article 0002a, posted 26 June 1998.

36. P. Petrovic, E. Kalso, K. M. Petersson, and M. Ingvar, Placebo and opioid analgesia—imaging a shared neuronal network, *Science* 295(5560): 1737–40 (2002).

37. A. Khan et al., Symptom reduction and suicide risk among patients treated with placebo in antipsychotic clinical trials: an analysis of the Food and Drug Administration database, *American Journal of Psychiatry* 158:1449–54 (2001); A. Khan et al., Symptom reduction and suicide risk among patients treated with placebo in antidepressant clinical trials: an analysis of the Food and Drug Administration database, *Archives of General Psychiatry* 57:311–17 (2000).

328   38. J. Dixit, New! improved! and still 100 percent fake, *Washington Post*, 19 May 2002.

39. One indirect type of test that suggests that antidepressants might beat placebos over a long time span are withdrawal studies. In these trials, after two groups A and B have both been given the active drug for some time, the patients in group A are switched to a placebo, and those in group B continue to receive the active drug group, but at reduced dosages. Generally, patients in the active drug group do better than those in the placebo group. However, this might be due to patients who switch to place-bos becoming unblinded (due to lack of side effects), rather than being in-dicative of a long-term effect of antidepressants over placebos.

40. P. G. Ney, C. Collins, and C. Spencer, Double blind: double talk or are there better ways to do research? *Medical Hypotheses* 21:119–26 (1986).

41. The published literature will probably become even more compro-mised given a recent editorial decision by the prestigious *New England Journal of Medicine* to allow physicians who receive consulting fees from drug companies to write review articles on new drugs.

42. Private e-mail communication, May 16, 2002. Here is Dr. Temple's full response:

> I have no precise numbers for you but we're aware that chance could explain a successful trial if enough studies were done. So, no, 2/20 will not get you approval. Note, though, that in depression, most trials are 3-arm (new drug, control active, plbo). A trial that can't show the effect of an active drug is a null result—it provides no information on the effective-ness of the new drug. Such a trial that showed the effect of the control but no effect of the new drug would be a very negative outcome. If there are only two arms you can't tell whether the study is the problem or the drug is the problem. I suppose if you had 18 no test studies and only 2 that had assay sensitivity (ability to tell active drug from placebo) one might con-sider that adequate, but it has never happened. In practice for sympto-matic conditions, there are usually more than two successful trials.
>
> It is unusual to incorporate into placebo something that mimics a drug's side effects. The few studies on this issue suggest that unblinding is not as easy or extensive as might be supposed.
>
> In most cases there is no requirement for a specified effect size unless loss of effect would be dangerous (e.g., for antibiotics we usually ask for evidence that loss of more than say 10 percent of the cure rate has been ruled out. Similarly, beta agonists are asked to improve FEV1 by at least 15 percent. For many symptomatic conditions, tho, being superior to plbo is the standard (a standard incompatible, I note, with being no bet-

329   ter than placebo). For outcome studies (mortality, etc.), being superior to placebo is the standard unless there is existing therapy in which case comparative trials are generally done.

Hope that answers the questions. As I said, drugs that are accepted for approval are not placebos.

43. Private e-mail communication with Dr. Robert Karlson.

44. D. P. Phillips et al., The hound of the Baskervilles effect: a natural experiment on the influence of psychological stress on the timing of death, *British Medical Journal* 323:1443–46 (2001).

45. R. Voelker, Nocebos contribute to a host of ills, *Journal of the American Medical Association* 275(5): 345–47 (1996).

46. W. B. Cannon, Voodoo death, *American Anthropologist* 33 (1942).

47. R. Ehrlich, *Nine Crazy Ideas in Science: A Few Might Even Be True* (Princeton, NJ: Princeton University Press, 2001).

48. Apart from the deaths in the immediate aftermath of Chernobyl, the estimates of the long-term cancer deaths vary widely, with some estimates ranging upward to nearly half a million, but others being much lower. In all likelihood the true number of cancer deaths may never be known, since the radiation-induced cancers will probably not cause a sufficiently large increase over the naturally occurring ones during the next half-century. One of the most feared radiation effects in the public imagination is probably genetic damage, and the creation of babies having monsterous defects. Genetic defects have been observed in animal experiments, and they are no different than the types of defects that occur spontaneously. For humans, even at Hiroshima and Nagasaki, where radiation levels were extremely high, the number of genetically defective babies born to exposed mothers was insufficiently higher than background to say how many had actually occurred.

49. M. Fumento, How the media and lawyers stir up false illnesses, at overlawyered.com/articles/fumento/nocebo.html.

50. R. Ader and N. Cohen, Behaviorally conditioned immunosuppression, *Psychosomatic Medicine* 37(4): 333–40 (1975).

51. F. Quitkin, Placebos, drug effects, and study design: a clinician's guide, *The American Journal of Psychiatry* 156(6): 829–36 (1999).

52. Khan et al., Severity of depression and response to antidepressants.

53. See Fischer and Greenberg, *The Limits of Biological Treatments*, 311.

54. R. P. Greenberg, R. F. Bornstein, M. D. Greenberg, et al., A meta-analysis of antidepressant outcome under "blinder" conditions, *Journal of Consulting and Clinical Psychology* 60:664–69 (1992); R. P. Greenberg, R. F. Bornstein, M. J. Zborowski, et al., A meta-analysis of fluoxetine outcome

330 in the treatment of depression, *Journal of Nervous and Mental Disease* 182: 547–51 (1994).

55. J.A.C. Sterne and G. Davey Smith, Sifting the evidence: what's wrong with significance tests, *British Medical Journal* 322:226–31 (2001).

Chapter Nine. Should You Worry about Your Cholesterol?

1. U. Ravnskov, *The Cholesterol Myths: Exposing the Fallacy That Saturated Fat and Cholesterol Cause Heart Disease* (Washington, DC: New Trends, 2000).

2. R. J. Simes, Low cholesterol and risk of non-coronary mortality, *Australian and New Zealand Journal of Medicine* 24(1): 113–19 (1994).

3. See easydiagnosis.com/articles/cholesterol4.html.

4. The main causes of excess deaths were cancer and stroke, and for women non-illness mortality. G. Lindberg, J. Merlo, L. Rastam, and A. Melander, Cause specific mortality at the lower extreme of serum cholesterol distribution: a population based cohort study (paper presented at the American Heart Association meeting, Anaheim, 11–14 November 2001).

5. D. R. Kaslow, *Cardiovascular Efficiency vs. Nutritional Deficiency* (San Diego, CA: International Foundation for Nutrition and Health, 1997); also, see www.drkaslow.com/html/cholesterol.htm.

6. Ravnskov, *The Cholesterol Myths*.

7. Mixed reviews have appeared in two minor journals: R. A. Riemersma, *European Journal of Lipid Science and Technology* 104:185 (2001); and J. M. Kauffman, *Journal of Scientific Exploration* 15:531–41 (2001); and the latter (favorable) one can also be found on the web at Amazon.com.

8. R. Ehrlich, *Nine Crazy Ideas in Science: A Few Might Even Be True* (Princeton, NJ: Princeton University Press, 2001).

9. K. Uraneck, No trials without money, no money without trials, *The Scientist* 16 (27 May 2002): 11.

10. The correlation coefficient was computed from the original data contained in J. Yerushalmy and H. E. Hillboe, Fat in the diet and mortality from heart disease, *New York State Journal of Medicine*, 15 July 1957, 234–54.

11. M. G. Marmot and S. L. Syme, Acculturation and coronary heart disease in Japanese-Americans, *American Journal of Epidemiology* 104(3): 225–47 (1976).

12. Ravnskov, *The Cholesterol Myths*.

13. Ibid.

14. D. Reed, D. McGee, J. Cohen, K. Yano, S. Syme, and M. Feinleib, Acculturation and coronary heart disease among Japanese men in Hawaii, *American Journal of Epidemiology* 115:894–905 (1982).

15. Ehrlich, *Nine Crazy Ideas*.

331   16. S. L. Malhotra, Epidemiology of ischaemic heart disease in India with special reference to causation, *British Heart Journal* 29:895–905 (1967).

17. Ravnskov, *The Cholesterol Myths*.

18. For information on the Masai, see G. V. Mann, R. D. Shaffer, and H. H. Stanstead, Cardiovascular disease in the Masai, *Journal of Atherosclerosis Research* 4:289–312 (1964); G. V. Mann et al., Atherosclerosis in the Masai, *American Journal of Epidemiology* 95:26–37 (1972). For information on the Samburu, see A. G. Shaper, Cardiovascular studies in the Samburu tribe of northern Kenya, *American Heart Journal* 63:437–42 (1962).

19. J. Day et al., Anthropometric, physiological and biochemical differences between urban and rural Maasai, *Atherosclerosis* 22:149–92 (1975).

20. See studies cited in note 18.

21. J. Hirsch, quoted in G. Taubes, The epidemic that wasn't, *Science* 291:2540 (2001).

22. G. Taubes, The soft science of dietary fat, *Science* 291:2536–45 (2001).

23. U. Ravnskov, The questionable role of saturated fat and polyunsaturated fatty acids in cardiovascular disease, *Journal of Clinical Epidemiology* 51:443–60 (1998).

24. E. B. Ascherio et al., Dietary fat and risk of coronary heart disease in men: cohort follow up study in the United States, *British Medical Journal* 313:84–90 (1996). Also, see www.hsph.harvard.edu/nutritionsource/fats.html.

25. "Many of the problems with trans-fats have been known or suspected for 15 to 20 years, but have been largely ignored in the US. In Europe, trans-fats are restricted in food products, and some countries allow no more than 0.1 per cent trans-fatty acid content. In contrast, margarines in the US may contain up to 30 to 50 per cent! Of course, the food industry denies there is any problem with this." At www.karlloren.com/Diabetes/p44.htm.

26. See www.hsph.harvard.edu/reviews/transfats.html.

27. W. P. Castelli, Cholesterol and lipids in the risk of coronary artery disease—the Framingham Heart Study, *Canadian Journal of Cardiology* 4 (suppl. A): 5A–10A (1988).

28. L. J. Seman et al., Lipoprotein(a)-cholesterol and coronary heart disease in the Framingham heart study, *Clinical Chemistry* 45(7): 1039–46 (1999); W. P. Castelli, The triglyceride issue: a view from Framingham, *American Heart Journal* 112(2): 432–37 (1986).

29. Ascherio et al., Dietary fat and risk.

30. Taubes, The soft science of dietary fat.

31. Ibid.

332   32. Ibid.

33. Quoted in Taubes, The soft science of dietary fat.

34. Ravnskov, *The Cholesterol Myths*.

35. J. Stamler et al., Relationship of baseline serum cholesterol levels in 3 large cohorts of younger men to long-term coronary, cardiovascular, and all-cause mortality and to longevity, *Journal of the American Medical Association* 284(3): 311–18 (2000).

36. Ibid.

37. It could be argued that the seat belt analogy is inappropriate to coronary heart disease, because there is no risk associated with wearing seat belts, while for CHD there might be a risk of lowering cholesterol, if Ravnskov were correct. However, in fact, there may be a risk of wearing seatbelts in certain types of collisions, and some have advanced a theory that wearing seatbelts makes drivers less cautious.

38. For the PROCAM study, see P. Cullen, H. Schulte, and G. Assmann, The Munster Heart Study (PROCAM): total mortality in middle-aged men is increased at low total and LDL cholesterol concentrations in smokers but not in nonsmokers, *Circulation* 96(7): 2128–36 (1997); for the Chicago Heart Association Study, and the People's Gas Company Study, see Stamler et al., Relationship of baseline serum cholesterol levels; for the Framingham study, see Seman et al., Lipoprotein(a)-cholesterol and coronary heart disease in the Framingham study.

39. K. Anderson, W. Castelli, and D. Levy, Cholesterol and mortality: 30 years of follow-up from the Framingham study, *Journal of the American Medical Association* 257(16): 2176–80 (1987).

40. H. M. Krumholz et al., Lack of association between cholesterol and coronary heart disease mortality and morbidity and all-cause mortality in persons older than 70 years, *Journal of the American Medical Association* 272:1335–40 (1994).

41. T. Harris, E. F. Cook, W. B. Kannel, and L. Goldman, Proportional hazards analysis of risk factors for coronary heart disease in individuals aged 65 and older: the Framingham Heart Study, *Journal of the American Geriatrics Society* 36(11): 1023–28 (1988).

42. Ravnskov, *The Cholesterol Myths*.

43. National Institutes of Health, Third Report of the Expert Panel on Detection, Evaluation, and Treatment of High Blood Cholesterol in Adults (Adult Treatment Panel III) Full Report. See www.nhlbi.nih.gov/guidelines/cholesterol/atp3_rpt.htm.

44. Anderson, Castelli, and Levy, Cholesterol and mortality.

45. S. J. Sharp and S. J. Pocock, Time trends in serum cholesterol before cancer death, *Epidemiology* 8(2): 132–36 (1997).

333   46. They are age, sex, diabetes (Y/N), smoker (Y/N), systolic BP, diastolic BP, HDL, and LDL.

47. M. J. Emond and W. Zareba, Prognostic value of cholesterol in women of different ages, *Journal of Women's Health* 6(3): 295–307 (1997).

48. On second thought, throwing out your TV might actually have some positive effect on your heart, if you spent the time exercising instead of watching TV—though you might try exercising on a stationary bike that powered a TV!

49. Ascherio, Dietary fat and risk.

50. Ascherio, private e-mail communication.

51. Kannel, private e-mail communication.

52. L. E. Ramsay, W. W. Yeo, and P. R. Jackson, Dietary reductions of serum cholesterol concentration: time to think again, *British Medical Journal* 303:853–957 (1991).

53. D. D. Gorder, T. A. Dolecek, et al., Dietary intake in the Multiple Risk Factor Intervention Trial (MRFIT): nutrient and food group changes over 6 years, *Journal of the American Dietetic Association* 86:744–58 (1986).

54. J. A. Resch, N. Okabe, and K. Kimoto, Cerebral atherosclerosis, *Geriatrics*, November 1969, 111–32.

55. Ravnskov, *The Cholesterol Myth*.

56. C. B. Taylor et al., Atherosclerosis in rhesus monkeys, II: Arterial lesions associated with hypercholesterolemia induced by dietary fat and cholesterol, *Archives of Pathology* 74:16–34 (1962); K. T. Lee, J. Jarmolych, et al., Production of advanced coronary atherosclerosis, myocardial infarction and "sudden death" in swine, *Experimental and Molecular Pathology* 15: 170–90 (1971).

57. Ravnskov, *The Cholesterol Myths*.

58. Ibid.

59. M. Leinonen and P. Saikku, Evidence for infectious agents in cardiovascular disease and atherosclerosis, *The Lancet Infectious Diseases* 2(1): 11–17 (2002).

60. U. Ravnskov, Cholesterol-lowering trials in coronary heart disease: frequency of citation and outcome, *British Medical Journal* 305:15–19 (1992).

61. N. Qizilbash, Are risk factors for stroke and coronary diseases the same? *Current Opinion in Lipidology* 9(4): 325–28 (1998).

62. Ibid.

63. B. M. Meiser, Simvastatin decreases accelerated graft vessel disease after heart transplantation in an animal model, *Transplantation Proceedings* 25:2077–79 (1993); also S. Wilson et al., Simvastatin preserves the structure of coronary adventitial vasa vasorum in experimental hypercholesterolemia independent of lipid lowering, *Circulation* 105(4): 415–18 (2002).

334   64. P. Libby, P. Ridker, and A. Maseri, Inflammation and atherosclerosis, *Circulation* 105(9): 1135–43 (2002).

65. M. Hayden, M. Pignone, C. Phillips, and C. Mulrow, Aspirin for the primary prevention of cardiovascular events: a summary of the evidence for the U.S. Preventive Services Task Force, *Annals of Internal Medicine* 136:16172 (2002). See also www.ahcpr.gov/clinic/3rduspstf/aspirin/asprr.htm.

66. Ibid.

67. M. F. Muldoon, S. M. Manuck, and K. M. Mathews, Lowering cholesterol concentrations and mortality: a quantitative review of primary prevention trials, *British Medical Journal* 301:309–14 (1990). Also, see M. H. Criqui, Cholesterol, primary and secondary prevention, and all-cause mortality, *Annals of Internal Medicine* 115:973–76 (1991); G. D. Smith, and J. Pekkanen, Should there be a moratorium on the use of cholesterol lowering drugs? *British Medical Journal* 304:431–34 (1992); G. D. Smith, F. Song, and T. Sheldon, Cholesterol lowering and mortality: the importance of considering initial level of risk, *British Medical Journal* 306:1367–73 (1993); M. Muldoon et al., Cholesterol reduction and non-illness mortality: meta-analysis of randomized clinical trials, *British Medical Journal* 322:11–15 (2001).

68. J. C. LaRosa, J. He, and S. Vupputuri, Effect of statins on risk of coronary disease: a meta-analysis of randomized controlled trials, *Journal of the American Medical Association* 282(24): 2340–46 (1999).

69. F. M. Sacks et al., The effect of pravastatin on coronary events after myocardial infarction in patients with average cholesterol levels, *New England Journal of Medicine* 335:1001–9 (1996).

70. J. A. Cauley et al., Abstract 1647, *Proceedings of the American Society of Clinical Oncology* (2001); also, see www.orlandosentinel.com/features/health/sns-health-breastcancer.story.

71. L. M. Bjerre and J. LeLorier, Do statins cause cancer? A meta-analysis of large randomized trials, *American Journal of Medicine* 110(9): 716–223 (2001).

72. T. B. Newman and S. B. Hulley, Carcinogenicity of lipid-lowering drugs, *Journal of the American Medical Association* 275:55–60 (1996).

73. Ibid.

74. Ibid.

75. Clofibrate was second on the list to occupational exposure to the fumigant ethylene dibromide. See L. S. Gold, T. H. Slone, B. R. Stern, N. B. Manley, and B. N. Ames, Rodent carcinogens: setting priorities, *Science* 258:261–66 (1992).

335   76. Newman and Hulley, Carcinogenicity of lipid-lowering drugs.

77. Ibid.

78. One of the statin drugs, Baycol (cerivastatin), was pulled from the market in 2001, because it was linked to 31 deaths. Apparently, the life-threatening side effects of Baycol were found to be more common among patients taking gemfibrozil simultaneously, but that drug remains on the market.

Chapter Ten. Epilogue

1. J. Taubes, What if it's all been a big fat lie? *New York Times*, 7 July 2002.

2. J. Groopman, Hormones for men: is male menopause a question of medicine or marketing? *The New Yorker*, 29 July 2002.

3. J. A. Sterne and G. Davey Smith, Sifting the Evidence: what's wrong with significance tests? *British Medical Journal* 322:226–31 (2001).

4. R. S. Root-Bernstein and S. J. Merrill, The necessity of cofactors in the pathogenesis of AIDS: a mathematical model, *Journal of Theoretical Biology* 187:135–46 (1997).

5. M. Alcubierre, *Classical and Quantum Gravity* 11:L73–L77 (1994).

6. For a description of the experiments and why they do not permit information transmission at FTL speeds, see www.plus.maths.org/issue12/news/fasterThanLight/.

7. P. J. Steinhardt and N. Turok, A cyclic model of the universe, *Science* 296(5572): 1436–39 (2002).

8. See www.usatoday.com/news/science/wonderquest/2001-06-20-time-travel.htm. Also, see the following scientific publication: R. L. Mallett, Weak gravitational field of the electromagnetic radiation in a ring laser, *Physics Letters* A 269:214 (2000).

9. W. B. Grant, An estimate of premature cancer mortality in the United States due to inadequate doses of solar ultraviolet-B radiation, *Cancer* 94(6): 1867–75 (2002).

Index